Nathaniel Alexander Nicholson

The Science of Exchanges

Fourth Edition

Nathaniel Alexander Nicholson

The Science of Exchanges
Fourth Edition

ISBN/EAN: 9783337035143

Printed in Europe, USA, Canada, Australia, Japan

Cover: Foto ©Suzi / pixelio.de

More available books at **www.hansebooks.com**

THE

SCIENCE OF EXCHANGES.

BY

N. A. NICHOLSON, M.A.,

TRINITY COLLEGE, OXFORD.

"Demand is in reality the bond which keeps all commercial dealings together. For if men wanted nothing or not so much, there would not be any or not so much commerce."—ARISTOTLE'S ETHICS.

FOURTH EDITION.

REVISED AND ENLARGED.

CASSELL, PETTER AND GALPIN,
LONDON, PARIS AND NEW YORK.

1873.

PRINTED BY WERTHEIMER, LEA AND CO.,
FINSBURY CIRCUS.

CONTENTS.

Chapter	Page
I.—Value	1
II.—Labour	13
III.—Buying and Selling	28
IV.—Currency	31
V.—Gold	72
VI.—The Foreign Exchanges	115
VII.—Depreciation of the Currency	140
VIII.—Land	155
IX.—Taxation	174
Conclusion	215

INDEX.

Abundance, 34
Absenteeism, 323
Accommodation bills, 108
Assay of gold, 208
Auxiliary currency, 96

Balance of trade, 78
Banker, 112
Banking, 87
Bank note, 113
Bill of exchange, 102, 106, 206
Budget, 326
Buying, 73

Capital, 49
Cheque, 114
Circulation, 90
Clearing house, 147
Commodity, 3
Comparative exchanges, 209
Competition, 20
Connexion in value between gold and silver, 187
Consumption, 6, 7, 8
Convertibility, 94
Course of exchange, 203
Credit, 81
Currency, 88
Customs, 281

Demand, 14
Deposits, 115
Depreciation, 218
Discount, 104
Double standard of value, 177
Duty on plate, 189

English coinage, 198
Exchange, 2
Excise, 280
Exchequer bills, 309
External panic, 119

Foreign coinages, 199
Free Trade, 37—320

Glut, 41

Hoarding, 52

Interest, 44
Internal panic, 120
Issue, 86

Journey, 198

Labour, 57
Land, 238
Light gold coin, 198

Millièmes, 189
Mintage, 208
Mint price, 198
Money, 83
Monopoly, 40

Panic, 118
Par (true), 200
Par (actual), 208
Price, 42
Production, 5
Profit, 10
Property, 48
Protection, 39—321
Pyx, 207

Rate of exchange, 201
Regulation of the exchanges, 217
Rent, 250
Reserve, 137
Retrenchment, 54

Saving, 53
Security, 109
Seignorage, 208
Selling, 74
Silver coinage, 199
Single standard of value, 199
Sources of wealth, 46
Specie point, 210
Speculation, 75
Standard of value, 188
Standard gold, 199
Strike, 67
Supply, 15

Tariff, 290
Tax, 258
Tendency of bullion, 209
Theory of value, 33
Theory of labour, 70
Theory of capital, 72
Theory of the coinage, 198
Tokens, 180
Trade, 36

Value, 1
Value of capital, 50
Value in exchange, 4
Valuing of gold, 207

Wages, 58
Want, 13
Wealth, 35
Weekly return of the Bank of England, 137

THE SCIENCE OF EXCHANGES.

CHAPTER I.

VALUE.

1. What is value?

Value is the real or fancied advantage which men derive from the use of anything, whether it be an exchangeable article, or not.

2. What is exchange?

Exchange is the giving something which has a value in order to get something in return which has also a value.

3. What is a commodity?

A commodity is anything which has a value and can be exchanged.

4. What is value in exchange?

It is the quantity of any other commodity which we can get in exchange for the particular quantity of the commodity which we give. When the word value is used in the science of which we treat, we always mean value in exchange, unless we particularly specify that we are using the word in its first and simplest sense.

5. What is production?

It is turning the gifts of the earth into commodities.

6. What is consumption?

It is the using of commodities.

7. What is consumption for the mere sake of consumption?

It is the using of commodities for the purpose of enjoyment.

8. What is consumption for the sake of production?

It is the using of commodities which is unavoidably caused by that production; it is the cost of production itself.

9. What is profit?

Any gain or advantage, from whatsoever source it is derived.

10. What is profit on a commodity?

It is the clear gain which any one makes by the exchange of that commodity.

11. What is the minimum of profit?

It is the lowest rate of profit which will satisfy the producers of any commodity, for without this minimum they would not continue to supply the commodity.

12. What brings the profit made by the producers of any commodity to its minimum?

Competition amongst each other for other commodities, for one undersells the other till the minimum of profit is reached.

13. What is want?

Want is the feeling that it would be to our advantage to have some commodity which we have not.

14. What is demand?

Demand is the offering a commodity which we have in exchange for one which we want.

15. What is supply?

Supply is the commodity itself which we offer in exchange.

16. What is effective demand?

It is offering a sufficient quantity of the commodity which we have to enable us to get the commodity which we want.

17. Can demand, as we define it, exist without supply, or supply without demand?

Certainly not, for we must furnish a supply of some sort or other, in order to offer it for the commodity which we want.

18. Does the furnishing a supply of any commodity create a demand for that commodity?

No, it does not; it only creates a demand for those commodities which the furnishers of that supply may want. Unless people offer commodities in exchange for the commodity supplied, there is no demand for it, and the supply remains on hand.

19. What does create a demand for a commodity?

The offering other commodities in exchange for that commodity.

20. What is competition?

It is offering more of any commodity, or a better quality of commodity, than others offer, in order to induce the suppliers of the commodity which you want, to deal with you instead of with the other offerers.

21. What effect has competition upon the value of any commodity competed for?

It always increases the value of that commodity, for the competitors, in bidding against each other, offer more of the commodities which they have, in exchange for the particular quantity of the commodity which they want, and thus increase its value.

22. What causes a diminution in the value of any commodity?

Competition for other commodities amongst those who supply that commodity, for they must offer more of it in exchange for particular quantities of those commodities which they are competing for.

23. What is the meaning of proportion between demand and supply?

The relative difference between the quantity of a commodity which is actually supplied in a particular market and the quantity which people want and offer for in that market.

24. When is a supply said to be too large?

When the quantity of a commodity actually supplied is larger than the quantity offered for by those who want the commodity.

25. When is a supply said to be too small?

When the quantity of a commodity actually supplied is smaller than the quantity offered for.

26. When is a supply said to be sufficient?

When the quantity of a commodity which is actually supplied is about equal to the quantity which is offered for.

27. What is the effect in each case on the value of the commodity supplied?

(1) When the supply is too large, those who supply the commodity must compete against each other for any commodities which they want, and thus diminish the value of their supply.

(2) When the supply is too small, those who want the commodity supplied must compete against each other to obtain it, and thus the value of that supply is increased.

(3) When the supply is sufficient, those who supply the commodity compete against each other for the commodities which they want, till they bring the value of their supply down to the cost of its production, and the minimum rate of profit to be made on its exchange.

28. What is average value?

The average value of any commodity to those who produce it consists of the average cost of its production and the average rate of profit to be made on its exchange; to the world at large it consists only of the cost of production, for this includes the average rate of profit which must be made by the producers.

29. What are fluctuations in value?

They are alterations in the proportion which the supply of any commodity in some particular market bears to the demand for it there.

30. What may be called the equilibrium of any commodity's value?

The average value of that commodity, for the commodity will rise above its average value or fall below it, according to the proportion which the supply of it in any particular market bears to the demand for it there.

31. What is intensity of demand?

The demand for any commodity is said to be intense when it much exceeds the supply of that commodity, and increases in intensity as long as people continue to offer more of the commodities which they have, in exchange for the particular quantity of the commodity which they want.

32. Does the demand for any commodity necessarily produce in process of time a supply proportionate to that demand?

It does not necessarily do so, for demand is merely offering a commodity which we have, in exchange for one which we want; intense demand for any commodity will certainly set many people at work to endeavour to supply it, for the profit to be made on its exchange will tempt them, but they may find great difficulty in procuring it themselves, from a variety of causes, and the supply may still be small and limited. In this case the only effect of intense demand will be to increase the value of the commodity; but in the general run of instances intense demand does lead to a proportionate supply, for all who can supply the commodity are tempted by the high rate of profit to do so.

33. Give a short theory of value.

If we define want to be the feeling that it would be to our advantage to have some commodity which we have not, we come at once into the presence of demand, or the offering a commodity which we have in exchange for one which we want; and we thus bring in our supply, or the commodity itself which we offer in exchange. Effective demand is the offering a sufficient quantity of the commodity which we have to enable us to get the commodity which we want, and the whole theory of value turns upon what makes the quantity, which we offer, sufficient to obtain the quantity which we want. To understand this we must define value in its two senses. The first sense is unconnected with exchange altogether. Value is the real or fancied advantage which men derive from the use of any thing, whether it be an exchangeable article or not. Value in its second sense is the quantity of any other commodity which we can get in exchange for the particular quantity of the commodity which we give. If any article had no value in the first sense of the word, it would assuredly possess no value in the second, for no one would exchange a thing they could make some use of for a thing they could make no use of at all.

Whatever be the exchange we may consider, we shall find that we do, in reality, exchange the real or fancied advantage which we derive from the use of the commodity which we have, for the real or fancied advantage which we hope to derive from the use of the commodity which we want, and are offering for.

Hence effective demand must always depend upon the sacrifice which the suppliers on each side are prepared to make in order to gratify their wants.

Mr. Thornton has done good service by drawing the attention of the student to this fact.

The tallow speculation a few years ago is a good instance to bring forward. Some merchants bought up all the supplies of tallow at a time it was rather scarce and dear, in order to command the market and drive the price up; but the speculation failed completely because the public were not prepared to make the sacrifice necessary to acquire the tallow under the circumstances, and they determined to do without tallow rather than pay the price demanded.

This brings us to the conclusion, that in particular cases the adjustment of the price must be in the nature of an experiment. Demand and supply may both exist without any actual transaction taking place, but once demand becomes effective, then an exchange takes place; hence effective demand for a commodity need not necessarily coincide with the supply of it in any market because but a few buyers may be willing to make the sacrifice which sellers demand of them, and a portion of the supply only will change hands while the rest goes away unsold. If there are seventy thousand sheep sent to a fair for sale, the supply of sheep remains the same, viz., seventy thousand, whether there be forty or fifty thousand of them sold; the demand for sheep may be very good, and yet sellers may ask such high prices that the effective demand may not be sufficient to carry off the whole number.

In the ordinary run of cases, the proportion between demand and supply regulates the price of the commodities exchanged.

34. What is abundance?

It is great plenty of any commodity.

35. What is wealth?

Wealth is the power of obtaining commodities which the possession of one or many commodities gives; it does not lie in the mere actual possession of commodities, because from circumstances they may have little or no value; it lies in the power which the possession confers.

36. What is trade?

Trade is the steady and continued exchange of commodities, which arises from a simple exchange having been found advantageous to both exchangers.

37. What is free trade?

Free trade is trade open to all who desire to engage in it without any regulations to restrict competition amongst them, whatever be their race or country, in their dealings with each other. When Colbert, says Doctor Franklin, assembled some wise old merchants of France, and desired their advice and opinion how he could serve and promote commerce: their answer, after consultation, was in three words only, "laissez-nous faire."

38. What do you mean by regulations to restrict competition?

Any regulations in any particular country which either shut out certain competitors altogether who would willingly offer their commodities in the markets of that country, or which press unfairly on them in the shape of duties on their commodities, to the advantage of other competitors who supply the same kind of commodities, but have no such duties to pay.

39. What is protection?

It is restricting competition in trade for the special advantage of particular classes of exchangers in any one country.

40. What is a monopoly?

A monopoly is permitting only a favoured few to supply any particular commodity, so as to prevent any competition amongst those who supply that commodity. It differs from protection, for this last admits of a restricted competition—the protected producers competing against each other.

41. What is glut?

Glut is when any commodity falls below the cost of its production, from the superabundant supply of it in any particular market.

42. What is price?

It is the quantity of money which a man can get in exchange for a commodity.

43. What is average price?

It is the average price of a commodity during a given time.

44. What is interest?

It is the price paid for the use of money.

45. What is rent?

It is the price paid for the use of land, houses, &c.

46. What are the sources of wealth?

The sources of wealth are production and trade.

47. What two things are most conducive to the development of a nation's wealth?

Cheap power of production and free trade; that is, the being able to produce commodities cheaply, and the being allowed to compete without restrictions in any market whatever, either as buyers or sellers.

48. What is property?

Anything which legally belongs to any one.

49. What is capital?

Any commodity, or combination of commodities, which is being employed with a view to increasing our capabilities of production by the material profit derived from that employment.

50. What is the best indication of the average value of capital in any country?

The average market rate of interest in that country, for capital in the shape of money can be easily applied to any purpose of production.

51. What ought the rate of interest to depend upon in any country where there are no usury laws?

It ought to depend upon the proportion between the quantity of money actually in use in that country, and the demand for its employment there. In new countries like Australia, where gold is produced in large quantities, the rate of interest is high, because every one who has capital in the shape of gold prefers to have it in any other shape, in order to make the most they can out of it, and the gold is consequently exported at once. Hoarding money, if practised on a large scale, as it used to be in France, will have the effect of raising the average rate of interest.

52. What is hoarding?

Hoarding is the accumulation of unemployed commodities.

53. What is saving?

Saving is the accumulating of commodities with the purpose of converting that accumulation into capital.

54. What is retrenchment?

It is saving by diminishing consumption for the sake of enjoyment, and is only useful to a nation in so far as it secures a better distribution of capital; but as it diminishes consumption for the sake of enjoyment it injures trade.

55. What is the best kind of saving for a nation?

Saving without retrenchment; for then consumption for the sake of enjoyment is not interfered with, and the nation produces each year more than it can consume, and so adds to its wealth.

CHAPTER II.

LABOUR.

56. What is labour?

Any work, toil, or exertion.

57. What is labour when considered as a commodity?

Labour is any work which a man performs in order to get other commodities in exchange for that work.

58. What are wages?

Any commodity given to the labourer in exchange for the work which he does. Wages may be either in money or in kind.

59. Does labour, in the long run, make a solid return to the labourer?

It does, for a labourer could not continue to work if he did not get other commodities in exchange for the work which he does.

60. Can a labourer be his own employer?

Certainly; any one who employs his labour in producing commodities which he exchanges for others that he wants, unites the characters of employer and labourer in his own person.

61. What is productive labour?

It is labour which makes a solid return to both employer and labourer, whether the labourer is his own employer, or whether he has an employer distinct from himself.

62. What is unproductive labour?

It is labour which makes a solid return to the labourer only. Any labour for a continuance, whether we call it productive or unproductive, must make a solid return to the labourer, or he could not continue to work; but it need not necessarily make a solid return to the employer. It is then called unproductive, because it is so as far as the employer is concerned.

63. How, then, can we distinguish the productive labourer from the unproductive one?

The productive labourer may be his own employer; the unproductive labourer must have a distinct employer, for he must find some one who is willing to give commodities in exchange for labour which makes him no solid return.

64. Are unproductive labourers of advantage to the prosperity of a nation?

They are; for if there were only productive labourers to be found in a great industrious nation, capital would be created much faster than it could be consumed, and the motive to further production would be taken away. In Great Britain, as it is, what we want is an extension of our markets; if we can only get customers for our goods, we shall produce the goods fast enough. A nation in a really healthy state ought always to produce more than she can consume, and thus, by saving, add to her wealth; but this balance of production over consumption is never very large, for the individuals of which the nation is composed live better as they get richer, and so consume more than they did; they also increase in numbers.

65. What does the value of labour depend upon?

It depends upon the proportion which the quantity of labour in the market bears to the demand for its employment. If labour is plentiful, it will be cheap; if scarce, it will be dear.

66. What is every bargain between an employer and a labourer?

It is a simple bargain between a buyer and a seller; the employer tries to buy the commodity, labour, as cheap as he can; the labourer tries to sell his labour to the best advantage, by giving as little of it as he can in exchange for the employer's money.

67. What is a strike?

It is a contention between employers and labourers about the price of labour. The labourers refuse to work unless they get increased wages, and the employers refuse to give the extra sum. Each side stands out, and those who can stand out longest are the winners.

68. Are strikes unjust in principle?

No, there is no harm in men trying to get the best price for the commodity in which they deal, whether it be labour or money.

69. Are strikes injurious to the general interests of society?

Certainly; every one suffers by business being brought to a stand-still; the public cannot get what they want, the employers are greatly injured by their capital lying idle, and the labourers have to suffer great privations before they can induce the employers to yield; indeed, they have sometimes to return to their work on the old terms.

70. Give a short theory of labour.

The first thing to be done with the term labour is to bring it into the domain of the science of exchanges, by excluding all labour which is not itself the subject of an exchange for some other commodity.

Labour is any work which a man performs, in order to get other commodities in exchange for that work.

The labourer, then, is a seller of the commodity, labour; he brings it to the labour market to exchange it for what he wants, and he ought to try to get the best price he can for his commodity.

The peculiarity which distinguishes labour from all other commodities, is that it is a necessary ingredient in their production. There may be more or less labour employed in the production of any commodity, but there is always some of this ingredient required.

The theory of labour, which we seek to establish, is based upon this fact; if we once admit that labour is an ingredient necessary to the production of all commodities, we must also admit that the more commodities we consume in a fair manner without waste, as long as we pay honestly for them, the more demand we create in the great labour-market of the world, and the better the chance the labourer has of selling his special commodity at its highest market price. Free trade is the labourer's best friend, because it enables us all to lay out the commodities which we have got to the best advantage, in purchasing those which we want, and by thus encouraging the consumption of commodities, it increases the demand for what all labourers have to sell—their work.

Every labourer in the community is benefited by the consumption of commodities which is always going on, whether it be a consumption for the sake of enjoyment, or a consumption for the sake of production, and the more commodities any nation can consume for the sake of production, the more active its trade will be, and the better price its working men will get for their work; I may add, too, that consumption for the sake of production will soon lead to consumption for the sake of enjoyment; they are twin brethren acting and reacting on each other, and they are ever at work carrying off the supplies which human industry is getting ready for them.

Whatever is produced, says Jean Baptiste Say, is sooner or later consumed; the products of industry have only been produced for the purpose of being consumed, and when any product is finished and fit for the market, if it remains on hand awaiting a consumer, the portion of capital which it represents lies idle ; now as all capital can be employed in a reproductive way, and can thus become a source of profit to its possessor, every product of industry, which is not consumed, occasions a loss equal to the profit, or, if you prefer the expression, to the interest which its value, if put in the shape of capital, would bring in ("Traité d'Economie Politique," book 3, chap. I). Trade is the voluntary and continuous exchange of commodities which arises from the different wants of different people; profit is the gain which each people makes on its own exchanges; the more, then, that we can multiply voluntary exchanges, the more we shall increase our profits.

Now what tends to multiply voluntary exchanges?

I reply the most absolute freedom being given to each people to exchange what it has in the markets it likes best. Free trade is what gives this freedom. If, then, we wish for free trade in labour, we must impose no peculiar restrictions on its exchange; we must allow each labourer to decide for himself how many hours he will work each day; for if we once attempt to cramp our productive powers as a nation by limiting a day's labour by law to eight, or nine hours, we are then treating full-grown men as if they were women or children; and we are giving other nations an advantage over us, for they will leave their labourers perfectly free, and then they will beat us in our own markets because they can get labour, the necessary ingredient in all commodities, cheaper than we can.

If the Trades-Unions could succeed in permanently shortening the hours of labour, and fixing the exact time which each man is to work during the day, the result would be a diminished production of commodities, bringing with it diminished means of obtaining the ordinary comforts of life. The lower classes would be the first sufferers, and the whole community would in time be affected, because diminished production must lead to diminished consumption.

Happily there is no likelihood of success in this case, because the labourer in England is as free as the employer, nor will he long continue to be the slave of either a Trades-Union or a capitalist.

The value of labour is regulated exactly as the value of any other commodity is regulated, nor can the labourer in the long run obtain more for it than its current price in a free market; if he holds out for more, he drives the employer either to seek for foreign workmen, who will be satisfied with less wages, or to withdraw his capital from a business which does not admit of sufficient profits being made.

I give the following account of an importation of foreign workmen to show what Trades-unions may bring about by ordering a strike. It is taken from the Scotsman paper, 1870.

"In the State of Massachusetts, the making of shoes is a very extensive branch of industry, giving employment to some 60,000 men. The manufacturers have for a long time been carrying on a game with their workmen, who had formed themselves into a trade-union, with the appropriate title of the "Society of the Knights of St. Crispin"—the purpose of each party in the game being to win such an advantage over the other as would vest in the victor the right to fix the rate of wages and the scale of profits. The trade society was very well organised; its members adhered faithfully to each other; at the command of their leaders they resorted to intimidation and violence as means of gaining their ends; and until lately the manufacturers were generally compelled to yield to the dictation of their workmen. About two months ago, a manufacturer in the town of North Adams, finding that the market for shoes was declining, proposed to his workmen that they should consent to accept a deduction of 10 per cent. on their wages until trade should revive. The proposition was referred to the leaders of the Crispins, and they appointed a committee to wait on the manufacturer with this message:—'We do not believe that the state of trade is as you represent it: but if you will submit your books to our inspection, in order that we may see for ourselves what orders you have to fill, and what prices you are to receive for the goods, we will be able to satisfy ourselves, and will then decide on the acceptance or the rejection of your offer.' The manufacturer declined to comply with this demand; the union ordered a strike; and the workmen left the factory. The manufacturer shut up his factory, and set out for California, whence he returned in a month with seventy-five Chinese labourers, whom he had engaged to take the place of his strikers. An angry crowd of the Knights of St. Crispin met the Asiatics at the railway station, hooted them, hustled them about, and threw stones at them; but they reached the factory safely. The manufacturer then prepared for defence; armed himself and his associates with 'six-barrelled revolvers,' and caused the newspapers to state that the Chinese were also armed with pistols, and with 'long knives that they could use with fearful dexterity;' and that they would be in no wise backward in defending themselves should they be attacked. The Chinese displayed the usual quickness of their race in learning the processes of their new business, and mastered the mysteries of the machines they were to use in a manner that promises full success to the experiment. They are under a contract to work for the manufacturer for the term of three years; they are to have twenty-three dollars a month for the first year, and twenty-six dollars a month for the rest of the time; they are to be furnished with lodgings, and with a certain quantity of food; and, if they die, their bodies are to be sent to China. The other shoe-manufacturers in Massachusetts are about to follow the example of the North Adams firm; two hundred and thirty other Chinese have arrived, and have been taken to different factories; and the question that now presents itself to the Crispins, and to the other trades-unions in Massachusetts, is what can be done to resist this invasion, which, if allowed to go on, will ruin their organisations, and place them at the mercy of employers? Thus far they have done nothing but hold meetings, at which they passed resolutions to carry the question into the coming elections, and to refuse to vote for any candidate who would not pledge himself to legislate against the importation of Chinese labourers into the country. It may be added that Congress seems to have been somewhat affected by the protests of the Crispins, inasmuch as it has incorporated into a bill for the amendment of the naturalisation laws a provision excluding Chinese from the rights of citizenship. This, however, can scarcely be expected to discourage Chinese emigration, for the Chinese who go to the United States are not animated at all by a desire to exercise political rights, but simply by a wish to earn money. The trades-unions, while insisting that the right to combine for higher wages is indisputable and sacred when exercised by themselves, demand that this same right shall not be exercised by the Chinese. The latter, who work for a penny a day at home, come to the United States in the hope of getting higher wages. Employers are willing to give them wages which are high to them, on condition that a large number of them will engage to work together, and for a certain period of time. In the opinion of the American trades-unionists, this is 'slavery,' and must be put down, by legislation if possible, and by violence if necessary."

I may add that the report of the Massachusetts bureau of labour, 1872, throws no light on the results obtained from this experiment. Mr. Sampson, the manufacturer referred to, positively refused to admit the Labour Commissioner into his shoe factory, or to give him any information whatever respecting his agreement with the Chinese.

The National Labour Union held a mass meeting at Albany to consider the question of the Chinese immigration, and one of their speakers put the matter very clearly before the meeting, thus :—

"When manufacturers consent to let the working man buy his goods in the open markets of the world, I, as a working man, will admit his right—mine having been admitted. What I object to is, to the paying of his prices, and the working at his prices too. If he has the right to buy his labour in the lowest market, I have an equal right to buy my goods in the lowest market. Let him try the experiment of competing with foreign goods, and I am willing to compete with foreign labour. When he consents to Free-trade, I'll consent to his importing his workmen; but to compel me to buy his goods at his rates, while he refuses to buy my labour at my rates, is not justice, and no arguing can make justice of it."

As to the second safeguard of the employer, viz :—the withdrawal of his capital, Mr. Brassey says "that our workmen are not sufficiently alive to the necessity which exists, for the utmost effort and ingenuity to enable capital invested in England to hold its own in the industrial campaign."

The Trades-Unions might be of real service to the working man if they would put the truth before him here and supply him with cheap and accurate information as to the state of trade.

Let us take the one article, iron. The advance in the price of iron in England is due to the increased cost of production caused by an advance in wages, a reduction of the hours of labour, and an increase in the price of coal; one of our ironmasters stated lately that he had to pay £32,000 for the same quantity of coke, which he used to procure for £12,000 a year ago. The cost of producing iron has been increased by about 40 per cent. in England. If this state of things continues, many orders for iron goods, which would formerly have been placed in England, will now be executed in other countries, where the work can be done for less money.

"Business in the iron trade in the north of England, though there is no material yielding of prices, is much less firm in tone than it was a month ago. Orders are not so pressing, and there is little desire on the part of buyers to enter upon new contracts at the present rates if it can be avoided. A good deal of iron is being shipped from the Tees, the Tyne, and Hartlepool, and pigs in small parcels are being sent to Germany. The iron shipbuilding trade, which has been so important and flourishing a branch of industry in the North of England during the past two years, within which period a very large capital has been sunk in new plant, does not look promising for the winter. Last autumn, it will be remembered, any amount of work might have been had by the builders, and very extensive contracts were then entered into by partnerships in iron steamers, for the northern iron shipbuilders to construct their new tonnage, the last of which contracts have been wrought off recently. Several large steamers were launched on the northern rivers last week. But a good deal more work will have to be got through before the principal iron shipbuilding yards in the north will be slack. There are, however, few new orders to replace those on hand. In many building yards, when vessels are launched, keels for other new ships are not laid. By the end of the year, therefore, unless there is a great change in the prospects of this trade, business is likely to become extremely dull, and a large number of workmen employed on it will have to be paid off."—*Times*, 26th September, 1872.

A lesson or two of this sort will prove advantageous to us before the short time movement goes too far.

In this case, as in all others, freedom within the law for all is the best safeguard for all. "Capital kann nur dann Allen wirklichen Segen bringen wenn die Arbeit, die Quelle des Vermögens und des Capitals, für Alle frei ist." (Max Wirth.)

71. Is labour, when considered as a commodity, to be looked upon as capital?

Capital has been defined to be "any commodity, or combination of commodities, which is being employed with a view to increasing our capabilities of production by the material profit derived from that employment." Now, a labourer will always employ his labour with this view, for whether the labour which he sells is profitable or unprofitable to his employer, it must be profitable to himself, or he would not continue to supply the commodity; labour, therefore, is always capital to the labourer. But labour is not necessarily capital to the employer, for he may pay labourers who are engaged in works conducted without any view to profit, and this sort of labour is not capital to him, although it is capital to the labourers themselves. Productive labour is capital to both the employer and the labourer; unproductive labour is capital to the labourer only, for it is unproductive as far as the employer is concerned.

72. Give a short theory of capital.

We must first define property, if we wish to get a true idea of what capital is.

Property is that which legally belongs to any one. As employed in the science of exchanges, the term property is a creation of law whether written or unwritten. The savage's idea of property rests upon custom or unwritten law.

Capital, therefore, must always be property, though this last need not necessarily be capital. Port wine, laid in at a low price, and maturing in the cellar of a wine-merchant in order to fetch a higher one at some future date, is the man's property, but it is a part of his capital also; while wine of the same class, under the same circumstances maturing in the Duke of Devonshire's cellar, is the Duke's property, but not his capital.

Dr. Whately says that almost all the definitions of capital which he is acquainted with pointedly exclude knowledge and skill. The reason of this is, that political economists have taken unnecessary trouble in keeping labour and capital distinct from each other; whereas, if we once admit labour to be a commodity, we cannot help admitting "that it may be employed with a view to increasing our capabilities of production by the material profit derived from that employment," (see 49), and it then becomes capital to all intents and purposes.

Let us take the instance of a skilled labourer with nothing but his labour to depend upon. A doctor may get on an average £10 a day in fees; he is able to pay his way honestly and live comfortably on £5; what of the other £5 which he invests in profitable securities? Are not they capital, and must not such a man, if he keeps proper accounts, count his labour for a day at so much money? Again, if we count our merchant steamers as capital, are we to exclude the skilled labour of their crews, when we include the coals which keep them moving? "The distinction between capital and non-capital," says Mr. Mill, "does not lie in the kind of commodities, but in the mind of the capitalist—in his will to employ them for one purpose rather than for another; and all property, however ill-adapted in itself for the use of labourers, is a part of capital so soon as it, or the value to be received for it, is set apart for productive reinvestment." ("Principles of Political Economy," People's edition, page 35.)

A shilling may therefore be capital as well as a thousand pounds, and the street Arab, who scrapes together a shilling to buy a shoe-brush, is a capitalist, and may live to be the possessor of many thousand pounds from such a beginning. But if a shilling may be capital, that part of a day's labour which a shilling represents is capital also. The common objection to this doctrine is that the day's labour is not realised tangible property, like the shilling, until the day is over. A labourer going to work may fall down and break his arm, undoubtedly he may; but substitute a dozen of eggs for the shilling or the labour, and may not the woman who carries them to market fall down too and break them?

A commodity is that which has a value and can be exchanged, and any commodity whatever may be so used, that we must acknowledge it to be capital.

Capital must of course be accumulated by saving, but most economists have created a difficulty for themselves here by not distinguishing between saving and retrenchment. Saving is the accumulation of commodities with the purpose of turning that accumulation into capital, while retrenchment is saving by diminishing consumption for the sake of enjoyment. A really healthy creation of capital results from saving, and not from retrenching, which can only be useful in special cases, where a better distribution of existing capital is suddenly called for.

Everything, then, which is being employed with a view to increasing our capabilities of production by the material profit derived from that employment is capital, whether it is a productive instrument, or the produce of that instrument to be afterwards employed with a view to profit. The land which produces a crop, the plough which turns the soil, the labour which secures the produce, and the produce itself, if a material profit is to be derived from its employment, are all alike capital.

Now, as all capital is necessarily property, the measure of a nation's prosperity and its real strength lies in the proportion which its capital bears to its other property, because the greater this proportion is, the more the nation in question can produce, and the more it can consume also. The two best means of increasing capital are, first education, to enable the workmen to diminish the cost of producing commodities; and secondly, free trade, to enable them to dispose of their commodities to the best advantage.

CHAPTER III.

BUYING AND SELLING.

73. What is buying?

It is giving money in exchange for any other commodity.

74. What is selling?

It is giving any other commodity in exchange for money.

75. What is speculation?

It is buying any commodity which is liable to alteration of price, when the supply of it is large and the price low, with a view to selling it when the supply chances to be small and the price high.

76. What is speculative demand?

It is any demand created by speculators, and not caused by the wants of trade.

77. What is foreign trade?

It is the exchanging the surplus produce of our own country for the surplus produce of some other country; we want to be supplied with commodities which are produced both better and cheaper abroad than we can produce them at home.

78. When are the foreign exchanges said to be favourable to a nation?

They are said to be favourable when that nation sells more to other nations than she buys from them, and the balance which then becomes due to her is called the balance of trade.

79. When are the foreign exchanges said to be adverse to a nation?

They are said to be adverse when that nation buys more from other nations than she sells to them.

80. When are the foreign exchanges said to be at par?

They are said to be at par between two nations when the amount of what is bought by the one in their mutual dealings exactly balances what is sold by the other.

81. What is credit?

Credit is the being able to obtain another man's capital upon trust. It is the means by which capital is put into the hands of those members of the community who can turn it to the best account, but who, not being themselves possessors of capital, would have to remain idle without the assistance which credit gives to them. Credit, therefore, does not create capital, but it gives capital already in existence a greater value than it would have possessed, if those who wanted capital for various enterprises, could not obtain it upon trust.

Sans contredire au vieil adage, Plus cautionis in re quam in personâ, il est trop certain qu'un pays auquel la probité manquerait généralement, et qui serait destitué notamment de ce fier et moderne sentiment qu'on appelle l'honneur commercial, devrait renoncer à voir le crédit fleurir dans son sein.—

Rien ne donne une moins favorable idée, si l'on peut s'exprimer ainsi, de la bonne tenue morale d'un peuple que d'être obligé, dans toutes ses transactions, d'avoir toujours l'argent à la main. L'expérience le prouve : le crédit ne s'établit à demeure que dans une population dont le moral présente de la solidité, où la masse des emprunteurs est honnête et intelligente, où, enfin, la manie de thésauriser et d'enfouir, qui paralyse le capital, est remplacée par l'activité laborieuse qui cherche avant tout à le féconder. Probité, intelligence, travail et sécurité, telles sont en tout lieu et en tout temps les conditions du crédit.—BAUDRILLART (*Manuel d'Economie Politique.*)

82. What is the golden rule in commerce?

Buy in the cheapest and sell in the dearest markets; without free trade this is impossible.

CHAPTER IV.

CURRENCY.

83. What is money?

It is the common measure by which we value all other commodities.

84. What is the great advantage which money possesses over all other commodities?

That people who have money can always find people who want money and are willing to exchange commodities for it, whereas people who have other commodities very often cannot find customers for them.

85. Mention any other advantages.

Money is easily brought from one market to another, and has been divided at the Mint into those sizes of coin which are most convenient for the purpcses of exchange.

86. What is issue?

It is the power of creating money.

87. What is banking?

It is the duty of using and distributing that money after it is issued in a proper manner in the conduct of business.

88. What is the currency?

The currency is the money which is in use in any country, and consists in Great Britain of coin and bank-notes, issued under certain restrictions.

89. What is managing the currency?

Managing the currency in Great Britain is regulating the fluctuations in the amount of bank-notes which we issue, by the fluctuations in the amount of gold coin, or bullion which comes in and goes out of the country, according to the state of the foreign exchanges.

90. What is the circulation?

The circulation is that amount of currency which is in use in any country.

91. How ought the circulation to be managed?

It ought to be left free to manage itself, expanding or contracting, according to the wants of trade.

92. What is the cause of alteration in the value of the currency?

Any continuing alteration in the circulation, whether it expands or contracts.

93. What is monetary pressure?

It is that rise in the rate of interest which ought to accompany any considerable contraction of the circulation.

94. What is convertibility?

It is an exact coincidence in value between bank-notes and the coin or bullion they are issued against.

95. What is a period of full currency?

It is when the foreign exchanges continue for some time at par, and the amount of the circulation is not affected by the state of the foreign exchanges.

96. What is auxiliary currency?

Bills of exchange, promissory notes, cheques, and anything which serves to economise the circulation, are called auxiliary currency; *as they are not made by law to fluctuate in their amount as bank-notes are, they cannot maintain a precisely equal value with true currency.*

97. What is meant by the term "in circulation"?

Anything which is performing the functions of money is in circulation. Bank-notes, performing these functions, are equally in circulation whether they are in the hands of the public or the bank which issued them; bills of exchange, promissory notes, cheques, &c., are in circulation whenever they are actually enabling us to do with a less amount of currency than if they were not in existence.

98. What is meant by the term, "in active circulation"?

Any kind of currency is in active circulation while one person passes it as money to another in constant succession.

99. What does the efficiency of the circulation in any country depend on?

It depends on the state of credit in that country; if the state of credit is good, a much less amount of currency will be sufficient for the purposes of trade than if the state of credit is bad. When people lose confidence they begin to hoard their money, and thus they diminish the amount of currency in active circulation. Auxiliary currency, too, is based upon credit, so that its power of economising the circulation is weakened or destroyed.

100. Is a want of money the same thing as a want of capital?

Certainly not; a want of money is a want of a particular kind of capital, not a want of capital generally.

101. What is the principal point of difference between a bill of exchange and a promissory note?

A bill of exchange contains an order to pay, whereas a promissory note contains a promise to pay.

102. Give some account of the parties to a bill of exchange.

The party who draws the bill is called the drawer; the party to whom the order to pay is addressed is called the drawee till he gives his assent to the transaction by writing his name across the bill, when he is called the acceptor; the party to whom the bill is made payable is called the payee; the drawer and the payee may be the same person. When any holder of a bill writes his name on the back of that bill, in order to make over his interest in it to some other person, he is called an indorser. The drawee, on accepting a bill, either personally or by his recognised agent, becomes primarily liable for that bill; but if he refuses to accept the bill when presented to him, or cannot pay it when it becomes due, he is said to dishonour it, and the drawer and any indorsers there may be become immediately liable to the holder, on receiving notice from him that the bill which he holds is dishonoured, accompanied by a request to pay its amount. If the drawer of the bill which has been dishonoured pays its amount, all the indorsers are discharged from their liability to the holder by such payment; but if the last indorser is compelled to pay the amount of the bill, the drawer and the previous indorsers are not discharged by this payment, for the last indorser can sue them for the amount he has paid to the holder of the bill. A bill may be paid any number of times before it is due, and may be put into circulation again between each payment; but once it is paid by the acceptor on its becoming due, it cannot be put into circulation again, nor can any action be brought upon it.

103. What is discounting a bill?

Any broker or banker who buys a bill, deducting so much from its price as interest till the bill becomes due, is said to discount that bill, and the interest which he deducts is called discount.

104. How does discount differ from interest?

Discount is the allowance which is made in order to get a sum of money paid before it is due. It differs from interest in this particular that when we get a bill discounted, the discounter not only deducts a certain rate of interest, being the interest for the time the bill has to run on the exact sum which he pays to us, but he also charges us with interest at the same rate on this interest itself.

If the rate of discount is said to be five per cent., the discount on a bill for £105 to run for one year amounts to £5 5s. or £5 for interest on £100, and 5s. for interest on £5 and the present worth of such a bill is £99 15s.

105. What is money at call?

Money lent by bankers, with the option of having it back at any moment, to bill-brokers, on the security of bills which they deposit with the banks.

106. Give an instance to explain the use of bills of exchange.

The operation which takes place, is put very clearly by Mr. Mill, in the following passage:—
"A merchant in England, A, has exported English commodities, consigning them to his correspondent, B, in France. Another merchant in France, C, has exported French commodities, suppose of equivalent value, to a merchant, D, in England. It is evidently unnecessary that B, in France, should send money to A, in England, and that D, in England, should send an equal sum of money to C, in France. The one debt may be applied to the payment of the other, and the double cost and risk of carriage be thus saved. A draws a bill on B for the amount which B owes to him; D having an equal amount to pay in France, buys this bill from A and sends it to C, who, at the expiration of the number of days which the bill has to run, presents it to B for payment. Thus the debt due from France to England, and the debt due from England to France, are both paid without sending an ounce of gold or silver from one country to the other."—'Principles of Political Economy.'

107. What advantages do bills of exchange at short dates possess?

Though the drawer, acceptor, and indorsers of such bills may all of them be persons of doubtful credit, yet still the shortness of the date gives some security to the holder of the bill; though all of them may be likely to become bankrupts, it is a chance if they all become so in so short a time.

108. What are accommodation bills?

Inland bills, which are not based upon any trade transactions, either present or future, but which are drawn for the sole purpose of raising money. In general, where an accommodation bill is concerned, the only real creditor is the man who discounts the bill, and the only real debtor is the man who gets the amount of the bill minus that discount. It requires great experience to distinguish accommodation bills, when properly disguised by dishonest traders, from bills which really are based on trade transactions or are drawn in anticipation of trade transactions, which will certainly take place.

109. What is a security?

A security is that which is given in pledge for the performance of an engagement.

110. What does banking depend upon?

In all its forms it depends upon credit.

111. What is the duty of a banker?

He ought to act as an intermediate party between those who have money seeking employment, and those who have industry and enterprise, but want money to carry out their plans.

112. Define a banker?

A banker is a man who makes it his business to borrow and lend money already in existence.

The *Times*, in reviewing Mr. Hankey's book on banking, published in 1867, speaks of the substitution of accurate definitions for loose and shifting usage. The question: What is a pound? is then correctly answered, but when we come to the next question, What is a banker? though the answer is the correct one, it is not kept to throughout the review.

A banker, says the *Times*, is a man who makes it his business to borrow and lend money. As there is nothing here said about the creation of that money, it must be presumed to be "already in existence," and the addition of these three words would have prevented the able writer of the review from getting into the quicksands of loose and shifting usage.

A little further on we read, "there is no distinction in the nature of things between a banker's ordinary business (*i. e.*, the borrowing and lending money already in existence), and this (the issuing of bank-notes) which is denied him. If he receives a sum from a customer, he enters in his books his indebtedness to the customer, and he holds himself ready to repay the same to the customer on his order.

"Had he given in exchange for the deposit promises to pay on demand, he would have made a memorandum of the transaction as before; the difference being that instead of becoming indebted to a known person, he would be indebted to the person or persons from time to time holding the promises he had issued.

"It is a difference of machinery, not of the essence of the business."

We might almost despair of accuracy being obtainable in the science of currency, when we compare this with the following passage: "No one is constrained to become the creditor of a particular banker by opening an account with him, but an ordinary tradesman is almost compelled to become the creditor of the bankers of his neighbourhood by taking their bank-notes; unless he is prepared to see his business dwindle away, he must take the bank-notes which pass current round him." So the essence of the banking business is just the same whether bankers have voluntary customers, or are given by law the power of making their whole neighbourhood their creditors whether they like it or not.

Well does Mr. Stirling say in his practical considerations on Bank management, "A strange heresy has arisen that paper-money is not money." Bank-notes, or transferable promises to pay coin to bearer on demand, circulating side by side with coin in endless succession, liquidating debts like coin, and which, when in circulation, all business people are, as it were, compelled to take, are absolutely money; and people who complain of restriction in the issuing of them, might just as well complain that they have not the power of coining gold, silver, and bronze money.

Had the *Times* given a definition of a banker in accordance with its subsequent observations, it would have been in the following form :—

A banker is a man who makes it his business to borrow and lend money, and to give transferable promises made by himself to pay bearer coin on demand, in exchange for deposits, or in satisfaction of any claims made upon him.

This would be tantamount to proclaiming that a banker ought to be allowed to create money as well as to borrow and lend it.

113. What ought a bank-note to be?

It ought to be a transferable promise on the part of the British Government to pay to bearer on demand a certain amount of coined gold; circulating side by side with coin in endless succession; liquidating debts like coin; and a legal tender everywhere in the United Kingdom, except at certain fixed places where gold can always be obtained according to promise when required.

114. What is a cheque?

A cheque is an order directed to a banker to pay over a specified sum of money; it must be dated, be made payable to bearer, or order on demand, be signed by the drawer, and be duly stamped. " A crossed cheque," says Mr. Newman, in his admirable little work on bankers' cheques, " is a cheque payable either to bearer or order, having the name of a banker, or two transverse lines with the words 'and Co.' written across the face of it, and the effect of the crossing is to direct the banker on whom the cheque is drawn to pay the cheque to some other banker only. The crossing of a cheque is now deemed by law a material part of the instrument, and it is made a felony to obliterate it." Sir John Lubbock proposes to extend the advantages, which are certainly derived from the crossing of cheques, to Bank of England notes, by introducing a bill to enable them to be crossed also. I believe that such a proposal arises from a complete misapprehension on Sir John Lubbock's part as to the true nature of a bank-note. We shall next hear of a proposal to cross sovereigns themselves, to prevent them being stolen out of tradesmen's tills.

115. What are deposits?

Deposits are any public or private moneys entrusted to the care of bankers during business hours.

116. What effect has the deposit system upon the circulation?

It serves to economise the circulation, for bankers need only keep whatever reserve of bank-notes they find sufficient to meet the average demands of depositors; it also serves the same purpose to an enormous extent by making transfers from one account to another possible. The amount settled at the clearing house in London by transfer, during the year from 30th April, 1870, to 30th April, 71, was £4,018,464,000, and all this without the employment of a single bank-note or sovereign.

117. Can deposits be said to form a part of the circulation?

No, for the same bank-note might possibly be

lodged as a distinct deposit by two or three different depositors in the same week. Once a deposit is lodged in any bank, it becomes a debt owed by that bank; the amount of reserve in coin and bank-notes which banks must hold to meet the average demands of depositors, certainly does form a part of the circulation, but this is a widely different thing from the total amount of deposits lodged in the banks.

118. What is panic?

It is a state of general alarm, arising from loss of credit.

119. What is external panic?

External panic is a panic arising from the exhaustion of bullion, consequent upon continued adverse exchanges.

120. What is internal panic?

Internal panic is a panic arising from the amount of bank-notes issued against other security than that of bullion being too large, and thereby causing doubts as to the convertibility of the bank-notes and an internal demand for bullion.

121. What is the best guard against external panic?

The only true way of guarding against external panic is raising the rate of interest slowly and steadily during a drain of bullion, till at last it will become cheaper to export other commodities, and the foreign exchanges will turn in our favour. Whenever we buy more than we sell in our dealings with other nations the foreign exchanges become adverse, and the effect is the same whether we buy commodities or invest in foreign securities —only another form of buying.

122. Is there any other means of guarding against external panic?

The Bank of England may strengthen the reserve by gradually realising securities, but a sudden realisation of securities to a large amount is not to the advantage of the Bank or the public.

123. Is there any remedy for internal panic?

As banking is dependent upon credit, our monetary system gives us no remedy for an internal panic except the strict limitation of the amount of bank-notes which we allow by law to be issued against other security than that of bullion. Sir Robert Peel assumed that the circulation in Great Britain had never been reduced so low as seventeen millions, and he also assumed that it will never again be reduced lower than it has been reduced in former times, and he came to the conclusion that in the worst circumstances he might safely allow fourteen million pounds in bank-notes to be issued against other security than that of bullion, and still have three million pounds in gold, or thereabouts, necessarily remaining in the country. By his act of 1844 he divided the Bank of England into two departments, the issue department and the banking department, and he permitted the issue department to issue £14,000,000 in bank-notes against other security than that of bullion, any issue of bank-notes over this amount to necessitate a corresponding amount of bullion being received into the till of the issue department, that is, that any bank-notes issued over the amount of £14,000,000 should be only so many certificates of the deposit of a corresponding amount of bullion. Sir Robert Peel allowed the Bank of England, and it alone, to buy bullion with its own notes, and thus increase its issues. On bullion being withdrawn from the till of the issue department, a precisely equal amount of bank-notes must be cancelled by the Bank of England, so that the fluctuations in the amount of bank-notes issued over the fixed amount are regulated by the fluctuations in the amount of the bullion as it comes in and goes out of the country, according to the state of the foreign exchanges. Sir Robert Peel, at the same time, limited the issues of all the other existing banks in England to the exact average amount of bank-notes which each of them had in circulation during the space of twelve weeks preceding the 12th April, 1844; whatever this amount proved to be, no bank in England was to exceed it, whatever amount of gold they might hold in their tills. No London bank but the Bank of England can issue bank-notes, and no bank created after the 6th of May, 1844, can issue them either. The Bank of England at present issues £15,000,000 in bank-notes against other security than that of gold coin or bullion, for some of the private banks have ceased to exercise their right of issue, and the Act allows the Bank of England to issue two-thirds of the whole amount of bank-notes which these banks used to issue. In Scotland and Ireland bank-notes are not a legal tender;

in England, Bank of England notes are a legal tender everywhere except at the Bank of England and its branches; but there is this difference between the issues of the Scotch and Irish banks and the issues of the English banks (the Bank of England being excepted)—the latter cannot issue bank-notes above the fixed amount, but the former are allowed to issue over and above the fixed amount any amount of bank-notes they please, provided that for every bank-note so issued they possess a corresponding amount of coin, of which three-fourths must be gold, and one-fourth may be silver, either at their head offices, if they are Scotch banks, or at any of the four depôts chosen in Ireland for that purpose, if they are Irish. In Great Britain, then, the power of issue has been very strictly limited; but even with these limitations, if all holders of bank-notes were to present them simultaneously for payment, their demands could not be met; our monetary system must be dependent upon credit.

124. How does a bank increase its liabilities?

By increasing its securities, and thus paying out bank-notes or coin again when deposited in the course of business.

125. Can a bank fail as a banking concern and yet maintain the convertibility of its notes?

Certainly, if it invests too largely in securities, and those securities cannot be readily converted into cash.

126. How does a bank, which is strictly limited

in its power of issue, obtain the money to invest in securities?

By employing its customers' deposits and its own capital.

127. What makes it difficult to understand questions bearing on the issue of bank-notes?

Because people do not clearly separate in their minds the business of issuing from that of banking; issuing is creating money, banking is managing money after it has been issued.

128. What is the difference between paying bank-notes out of a bank in the ordinary course of business and re-issuing bank-notes?

Bank-notes deposited by customers in the ordinary course of business are already issued, and the bank pays them out again as so much money, merely keeping whatever amount is sufficient to meet the average demands made upon it; re-issuing bank-notes is issuing bank-notes again which have been tendered for payment in gold at the bank, and have been paid. The object of the Act of 1844 was to secure that over and above a certain fixed amount of bank-notes issued against other security than that of bullion, any additional issue of bank-notes should be only so many certificates of the deposit of a corresponding amount of bullion, and therefore, when bullion is withdrawn from the issue department, a precisely similar amount of bank-notes ought to be cancelled.

129. Why is the permission to issue a certain amount of bank-notes against other security than that of bullion so very valuable?

Because there is a double return of profit on capital thus invested: first, the dividends arising from capital invested in profitable securities; secondly, the interest payable on the bank-notes issued against those securities.

130. Is the power of issuing bank-notes over the fixed amount one which can be safely used by any other bank than the Bank of England?

It is very doubtful whether it is a safe power in the hands of the Scotch and Irish banks, even strictly limited as they are at present; they may, and perhaps often do, abuse the power granted to them.

131. Give an instance to show how the Scotch and Irish banks may avail themselves of the power of extra issue.

Let us suppose that one of those banks has an authorised circulation of £100,000, and notes issued to that amount against other security than that of coin, and that it holds £25,000 in coin, of which £15,000 are at its depôt or head office, and £10,000 in its own till; by the Act of 1844 it might issue, if it pleased, £25,000 in bank-notes, by transferring the £10,000 in coin, which it holds in its own till, to its depôt or head office, but this would leave it no fund in coin to maintain the convertibility of its authorised issue. It issues, we will say, £20,000 in bank-notes, above its fixed amount, and transfers £5,000 in coin to its head office. Its affairs now stand thus: it has a total circulation of £120,000 in bank-notes, and it holds £25,000 in coin; but as it has issued

£20,000 in bank-notes over its fixed amount, it holds, in reality, but £5,000 in coin, against which no extra notes have been issued to meet its authorised issue of £100,000, this is one twentieth.

132. In what, then, does the unsoundness of such a system consist?

In the necessity which it entails on any banks which follow it of evading the Act of 1844, though for very short intervals, should any pressure or difficulties arise. A part of the coin which the banks ought to keep at their depôts or head offices, by the Act of 1844, if they issue any bank-notes over their fixed amount, can be called upon for a very short time, to help out the small fund of coin which the banks keep in their own tills to maintain the convertibility of their authorised circulation, and against which they issue no extra notes. Temporary as such an evasion of the Act of 1844 is, it gives the banks time to realise securities and put their houses in order should there be signs of pressure.

133. Can the Scotch and Irish banks make a profit on their extra issues without evading the Act of 1844?

As long as the times are prosperous and there is no pressure, they can; they are able to keep a much larger note-circulation afloat, and to give much more accommodation in consequence of their extra issues. The profit is not made absolutely on the extra issues themselves, for there is the expense of issuing, the wear and tear of the notes, and a duty of seven shillings per cent. on every issue to be taken into account; but on the increased business which the extra issues bring to the bank. When monetary storms and pressure arise, the Act of 1844 must be temporarily evaded, and a risk of awakening an internal panic amongst their customers must be run.

134. Give an instance to show how large their indirect profit must be.

About eight years ago the Bank of France for some considerable time paid £16,000 a month in premiums, buying gold; this was, of course, in itself a dead loss, but the bank considered it would be much more than compensated by the increase in its business, as it was able to keep a much larger note-circulation afloat, and to give much more accommodation than it would otherwise have done.

135. What has been often the ruin of joint-stock banks?

Making enormous advances to directors without proper security, and alluring depositors by high rates of interest, in order to obtain the funds necessary for those advances.

136. Is there any real safeguard against commercial disasters?

No; the tide of trade must ebb and flow; adverse circumstances, such as famine, or war, or undue speculation, will come upon us whatever we do; *but the Act of 1844 enables us to meet them better, because it preserves the convertibility of our paper-issues, and prevents the calamity of internal panic being added to external panic.*

137. What is the reserve in the Bank of England?

It is the amount of bank-notes kept by the bank in the till of the banking-department, in order to meet the average demands of depositors, and all other demands whatever made on the bank. When a drain of bullion begins, as the bank cannot get at the notes which are in the hands of the public, it must come on the reserve, and cancel bank-notes taken from it as the bullion is being withdrawn from the issue-department. The bank must take care to protect the reserve in order to be able to meet the demands made on her, and when a drain of bullion is going on, the way to do this is by slowly and steadily raising the rate of interest as the bullion goes on diminishing, for money is in reality getting scarce, and more ought to be paid for its use. The bank may still further protect the reserve by realising as large an amount of securities as she can conveniently do without injury to herself, and without awakening alarm in the minds of the public.

The Balance-sheet of the two Departments appears weekly in the "Gazette" in the following form:—

ISSUE DEPARTMENT.

Liabilities.	£	Assets.	£
Notes issued	37,507,905	Government Debt	11,015,100
		Other Securities	3,984,900
		Gold Coin and Bullion	22,507,905
		Silver Bullion	
	£37,507,905		£37,507,905

BANKING DEPARTMENT.

Liabilities	£	Assets	£
Proprietors' Capital	14,553,000	Government Securities	13,356,411
Rest (Undivided Profits and Reserve Fund)	3,749,970	Other Securities	20,930,994
Public Deposits, including Exchequer, Savings' Banks, Commissioners of National Debt and Dividend Accounts	7,896,805	Notes (Reserve)	11,067,120
		Gold and Silver Coin	641,355
Other Deposits	19,333,700		
Seven Day and other Bills	462,405		
	£45,995,880		£45,995,880

Dated 5th September, 1872.

FRANK MAY, *Deputy Chief Cashier.*

The above Balance-sheet, if presented in the old form before the separation of the Departments took place, would appear thus:—

LIABILITIES.	£	ASSETS.	£
Circulation with Bank Post Bills	26,903,190	Securities	34,734,405
Public Deposits	7,896,805	Coin and Bullion	23,149,260
Private Deposits	19,333,700		
	£54,133,695		£57,883,665
Balance of Assets or Rest	3,749,970		
	£57,883,665		£57,883,665

The advantage which the new form of stating the account confers, is that we can readily ascertain the proportion which exists between the supply of bank-notes and coin in the Banking Department, and the immediate demand which may at any time arise to carry off this supply.

The proper management of the Bank turns upon an accurate estimate of this proportion, and not upon the actual amount of the reserve taken by itself without reference to the amount and nature of the deposits.

By the new form of account the Bank holds two distinct reserves, one against its issue of bank-notes, and the other against the deposits intrusted to its management; by the old form of account the Bank held but one reserve against its circulation of bank-notes and its deposits combined; this prevented the public from obtaining a true view of the Bank's position as a banking institution.

If we take the Bank return of the 31st May, 1866 (answer 147), and if we compare the old and new forms of stating the account, we find that by the old form the Bank held a reserve in gold-coin, and bullion of £11,878,775, against a liability, in circulation and deposits combined, of £53,218,117; and that by the new form the reserve in the Banking Department had fallen to below one million, while the deposits amounted to over twenty-six millions and a half.

"The separation of the Departments," says Mr. James Stirling in his "Practical Considerations on Banking," "was not a change in form only, or, as it has often been called, a mere matter of account; on the contrary, it was a change of the most vital nature, relieving the Bank at once and for ever from the perplexities which had originated in the twofold character of its management. It subjected the currency to the control of a self-acting mechanism, following the natural movement of the precious metals, while it left the Bank free to concentrate its attention on its banking-reserve, and to conduct its money-dealing business according to the ordinary influences of the market. On the one hand it secured the convertibility of the note, by basing on Government securities the portion held to be the practical minimum of circulation, and by causing the remainder to vary exactly as it would have done if metallic; on the other, it enabled the Bank to follow the movements of the money-market, undistracted by extraneous considerations. The Act of 1844 provided a legal basis of indisputable solidity for the issues of the Bank, while it left to the discretion of the directors the amount of reserve to be held against deposits.

138. What effect has a proper management of the reserve on the foreign exchanges?

A steady increase in the rate of interest will check the exportation of bullion, which is only exported because it happens to be the commodity which we can best spare at that time, and the contraction of the circulation caused by the drain of bullion, and the corresponding withdrawal of bank-notes from the reserve, will soon lead to a fall of prices, which will induce foreign nations to buy more largely in our markets, and thus turn the foreign exchanges in our favour.

139. Is it wise of the Bank of England to refuse to discount for the discount houses?

Yes; if the Bank of England were to discount for the discount houses, she would expose herself to the danger of their uniting in times of panic, presenting very large amounts of bills, and demanding their discount on the threat of stopping payment if they did not get it. Such a course of action would embarrass the bank, and might cause most serious mischief to the country. The discount houses are in the habit of taking money at call to an enormous amount, and whenever they have to pay a high rate of interest for it, they can afford to keep but small cash-balances idle. As long as the Bank of England discounted for them they discounted bills rather under the bank rate, in order to get the business into their hands, and if bad times came, and their cash was run out, they had a great mass of bills on hand ready to present to the bank for discount in a threatening manner. Now that the bank has wisely refused to discount for them, and money begins to show signs of scarcity, they are afraid to discount bills under the bank rate, and the bills, instead of being massed together in the hands of a phalanx of dealers, find their way to the Bank of England, the National Bank, or any other of the strong money-holders who are in a firmer position than the discount houses, and not likely to increase the difficulties of the situation by any very sudden requirements.

140. What was the only check on the issues of the Bank of England before the Act of 1819?

The common rule of the bank, to discount only good mercantile bills at sixty-one days' date, at the rate of 5 per cent. per annum, which chanced to be about the market rate of interest at the time, with her extra issues of bank-notes. Notes issued on these terms were within some sort of limits; that is, they were not so pregnant with mischief as bank-notes issued under no checks at all, as the country banks issued them.

141. What is the plain principle involved in the matter?

That a currency, when composed of bank-notes and coin ought to be made by law to fluctuate in its amount exactly as a currency composed of coin only would have fluctuated under similar circumstances. There is no other means of securing that the public should always be able to obtain sovereigns in exchange for bank-notes without any loss or risk whatever.

142. Can we prevent the temporary inconvenience which commerce in England suffers from when an advance in our rate of discount is caused by any sudden demand for money on the part of any nation who deals with us?

A certain amount of pressure and inconvenience is unavoidable, but by strictly adhering to the Act of 1844, and by steadily advancing the rate of interest as our circulation contracts, the matter adjusts itself. No nation can obtain more from us than we owe them, and the more our rate of interest advances the more money is attracted to us from every quarter, for more can be had for it with us than elsewhere, so that pressure and inconvenience from this source cannot last long.

143. What is the gold and silver coin kept for in the banking-department?

It is kept for the convenience of customers; no bank-notes are issued against this gold coin, for then the bank-notes and gold against which they are issued would be in circulation at the same time. This amount is, like the stock in a jeweller's shop, necessary to the business of a bank, and the holding it is a dead loss.

144. If the currency were purely metallic, would that be a safeguard against panic?

Certainly not; as long as the deposit system prevailed, and auxiliary currency was in existence, bankers would then, as now, hold only whatever amount of gold they found necessary to meet the average demands of their customers; and if any extraordinary demand was to take place, they could only meet it by realising securities, which, in times of panic, must always be done at a loss.

145. Are there any valid objections to three central issuing establishments for the whole of the United Kingdom, under the control of the House of Commons?

The objections do not appear to be valid. Objectors say that it would be dangerous to allow so large an amount of bank-notes to be issued against Government securities, and that it would render convertibility impossible in times of panic; but, from the very nature of our monetary system, convertibility must always be impossible in times of panic, for if all the holders of bank-notes were to present them for payment in gold at one time, their demands could not be met, the securities against which our limited amount of bank-notes are issued could not be converted into gold during such an emergency. As long as the amount of bank-notes permitted to be issued in the United Kingdom, against other security than that of bullion, is lower than the circulation could ever be reduced to in the worst possible circumstances, we are safer with our bank-notes issued against Government securities than we could be if they were issued against any other securities whatever other than bullion itself. Objectors also say that entrusting the business of issuing bank notes to private enterprise secures more care, diligence, and economy in the management of that business, than if it was entrusted to three central issuing establishments under the control of the House of Commons. The same might be said of any Government establishments whatever, but with less truth in this special instance, for the business of issuing bank-notes is purely mechanical, only requiring the very best preventives against the possibility of forgery that science can discover.

146. Sketch a plan for a new monetary system?

What we propose is, that all bank-notes above a certain amount, to be fixed by the House of Commons, should be issued by Government in exchange for properly assayed gold 22 carats fine at the rate of £3 17s. 10½d. per ounce, exactly as sovereigns are now issued by the Mint in exchange for the same kind of gold at the same rate. Giving bank-notes in exchange for bullion at the exact value of this last per standard ounce, and which is to lie idle in a vault, as long as the amount of bank-notes, which represents it, is in circulation, is a very different thing from being a dealer in bullion;

nor does it involve any transactions in either home or foreign securities.

The House of Commons must determine what amount of bank-notes may be issued against securities other than gold coin or bullion, and this amount should be unalterable without their consent. The Government, once this amount of bank-notes was issued, would have nothing to say to the various bankers in the country; its duty would be simply to take care that any one who wanted bank-notes in exchange for bullion or gold coin, should be able to get them, or *vice versâ*. This could be arranged by three issuing establishments at London, Edinburgh, and Dublin, where stores of bullion, gold coin, and bank-notes, should be kept for the purpose of being exchanged one for the other, according to the wants of the public. The bank-notes of the State should be a legal tender everywhere but at the issuing establishments; here they would be simple certificates, establishing bearer's claim to the quantity of gold they represented, and they should be allowed to go as low as one sovereign. This last regulation would tend to keep holders of bank-notes away from the issuing establishments until they required gold for exportation, and the tedious process of changing five-pound notes into sovereigns, and back again into five-pound notes, would be thereby avoided.

I have never seen any objection that I considered valid against one-pound notes when issued under proper regulations; and any one who has lived in Scotland or Ireland, and has given any attention to the subject, will admit their convenience of carriage and superiority to sovereigns, which are often light, and may have to be weighed.

The legitimate business of a banker is to borrow and lend money already in existence, and the issuing establishments referred to would in no way interfere with this business; it is their part to create the gold and bank-note currency, which all Her Majesty's subjects may then use to their best advantage.

To each of the three issuing establishments there should be a Government clearing-house department attached, which should be open to all bankers on payment of an annual fee; they should be permitted to settle with each other by drafts on the reserves which they had previously lodged at the clearing-house department, just as they now do by drafts on the Bank of England, such drafts to be payable to clearing bankers only.

This system would extend the convenience of the clearing system to Scotland and Ireland, and allow the Bank of England to assume her true position as bank to the State, and therefore the great joint-stock bank *par excellence*, and no more. I do not propose that the Government should keep their own accounts, manage the public debt, and collect the revenue; the Bank of England ought always to be the bankers of the State, and to receive a sufficient remuneration for being so.

Nothing would, with my system, be heard of duty on the part of the Bank of England to help the whole community in times of difficulty; and the principal advantages gained thereby would be that no bank-note would then circulate in the country, unless it was just as secure to the holder as a sovereign, and that dangerous speculations would be discouraged in their beginnings by the natural system of each bank holding its own separate reserve, and being thus forced to care for its own solvency.

147. Explain the operations of the clearing house?

The clearing house, says Mr. G. G. Newman, is a large room fitted with desks. Each banker using the house has one or more of these, marked with his name or firm. In the morning and at three o'clock in the afternoon of each week-day a clerk from each banker using the house attends, bringing with him the cheques on other banks that have been paid into his bank since last clearing; these he deposits on the desks of the respective banks on whom they are drawn. He then credits their accounts separately with the different amounts they have placed on his desk as against his bank. Balances are then struck from the several accounts, and a general balance of all the balances of the several accounts, debtor or creditor, as the case may be, is struck, in which, of course, the grand total of the amount to be received by the several banks agrees with that to be paid. Each bank then gives or receives, as the case may be, a ticket or draft on the Bank of England for the amount due to, or owed by it on the result of its own balance. Every bank using the clearing house must for the purpose keep an account at the Bank of England. (Summary of the Law relating to Cheques, 2nd edition, page 35).

We add the following extract from Mr. Laing's "Theory of Business," 2nd edition, page 171, to complete the picture :—

" Banks which, as the result of the day's operations are in the position of having to pay a balance to the clearing house, hand a peculiar white cheque on their account at the Bank of England, technically called a transfer ticket; whereas banks which have to receive a balance, pass a green ticket, instructing the Bank of England to place the amount specified on it to their credit. The transfer tickets are conveyed to the bankers' section of the Private Drawing Office, where the clearing house account is kept. The sum of the white and green tickets should exactly balance one another at the end of the day. The Bank of England has recently been admitted to the clearing, so far as to pass cheques on other bankers received from its customers through the house; other banks, however, still bring their cheques on the Bank of England to it at once for encashment. The effect of this arrangement on the condition of the clearing account is, of course, to place the Bank in the position of always holding a green transfer ticket, that is, of having to receive a balance."

The whole clearing house system is clearly explained in "London Banking," by Mr. Ernest Seyd. We refer the student to his little work, extracting the following *resumé* of the system:—

1. We have the cheque system by the customers.
2. Its first concentration from the customers into the hands of the bankers.
3. The reduction in the clearing house to balances between each of the bankers.
4. The further reduction of these balances into one final balance for each banker.
5. The absorption of the twenty-six final balances through the clearing house account at the Bank of England" (Page 55).

Now we advance nothing against the clearing house system *per se;* it is an admirable one for the convenience of bankers and their customers, and we should try to give bankers throughout the United Kingdom increased facilities for clearing, rather than to diminish the existing ones.

This is not the weak point of our present monetary system.

I assert that the Bank of England should cease to hold the working cash reserves now deposited there by all the clearing bankers.

These reserves may be looked upon by the Bank either in the light of deposits, or in the light of a trust to be held untouched for each bank; if they are looked upon as deposits and treated as such, the Bank makes a hazardous and therefore illegitimate profit on their use; if they are looked upon in the light of a trust, the Bank can make nothing by holding them for the banks, and ought not to be required to perform such a service.

In my opinion the clearing house ought to be made a government department, and a clearing office established in each of our three issuing establishments, previously referred to, with special vaults for the keeping of the cash reserves of all bankers who wish to enter this department as clearing bankers.

We economise ready money too much, because we are not satisfied with the simple economy of depositors entrusting their money to our banks, and these last holding a sufficient reserve in actual money to meet the demands of their depositors; but our bankers themselves must become in their turn depositors, and entrust the greater part of their reserves to the Bank of England which, as Mr. Hankey truly remarks, cannot be made to supply ready money beyond what, under the ordinary good management of a deposit bank, it can retain in reserve, that is to say, that while our present clearing house system continues as it is, the Bank of England is a deposit bank, well managed it is true, but still dealing with enormous banking establishments in London as depositors.

We extract the following passage from the "Times" (City Article, 29th May, 1869) as the best practical commentary on our observations:—
"The discussion last evening as to the probable effects of the new revenue arrangements on the money market has excited very little attention in the city.

Practical traders and acute speculators are alike conversant with the fact, that in whatever hands any portion of the currency may happen to be at any given time, it will not be allowed to lie idle, but will find its way to the discount market in order that it may yield some income to its possessors."

No language can be clearer or more true, nor could we cite a stronger confirmation of the views we put forward here.

The "Times" tells us that another panic may certainly be expected in 1876 ("Times," 31st May, 1869), and we tell the "Times" that, if our present system is allowed to continue, we shall certainly see a fourth suspension of the law on which our currency system is based.

Such a suspension ought not to be necesary, speaking theoretically as Economists are so fond of doing; but given the existing system, and looking at its practical working, it is necessary, for it is the only safety valve we have to prevent our business boilers being blown absolutely to pieces.

Commercial panics must occur from overtrading and speculation; what we suggest is not a means of preventing their occurrence, but of diminishing their effects when they do take place. We want to lop off the unsound branches, once their unsoundness is known, instead of leaving them where they are, to spread mischief through the whole tree. But no, this is too plain and simple a course to follow. The "Times" even declines to discuss it, it is quite impracticable.

Let us sacrifice everything to our great banking establishments; let us always stand in fear of ever-recurring panics; let all honest, regular traders throughout the United Kingdom suffer the dreadful infliction of a ten per cent. rate for three months when the great panic arrives.

What of this? Do not our bankers distribute dividends year after year, varying between fifteen and twenty per cent.; and ought not " rational combinations of powerful banking establishments " to be able to protect honest traders in their hour of need?

The panaceas hitherto offered are futile to the last degree; there is but one firm basis for the honest, regular trader to rest his foot upon.

We must make each bank hold its own reserve, stand on its own merits, and take care of its own solvency.

I here add, to illustrate my views, the return of the Bank of England for the week ending on Wednesday, 30th May, 1866:—

ISSUE DEPARTMENT.

Liabilities.	£	Assets.	£
Notes Issued	26,434,205	Government Debt	11,015,100
		Other Securities	3,984,900
		Gold Coin and Bullion	11,434,205
		Silver Bullion	
	£26,434,205		£26,434,205

BANKING DEPARTMENT.

Liabilities.	£	Assets.	£
Proprietors' Capital....	14,553,000	Government Securities	10,864,638
Rest	3,419,759	Other Securities	33,447,463
Public Deposits	6,188,512	Notes..............	415,410
Other Deposits........	20,467,080	Gold and Silver Coin ..	444,570
Seven Days' and Other Bills	543,730		
	£45,172,081		£45,172,081

Dated 31*st May*, 1866.

W. MILLER, *Chief Cashier*.

Or, if made out in the old form —

Circulation with Bank Post Bills	26,562,525	Securities	44,759,101
Public Deposits	6,188,512	Gold Coin and Bullion..	11,878,775
Private Deposits	20,467,080		
	£53,218,117		£56,637,876
Balance of Assets or Rest	3,419,759		
	£56,637,876		£56,637,876

The student may compare this return with that given in answer 137.

148. What is the advantage of publishing the account of the Bank of England every week?

It enables capitalists and men of business, by comparing the account of one week with another, to make some guess at what the value of money is likely to be. They look to the amount of bullion held by the bank, and the state of the reserve.

149. Would a statement of the same kind, but relating only to the central issuing-establishments proposed, have the same effect?

Certainly; capitalists and men of business would then only look to the amount of bullion held each week by those establishments.

150. What is the difference between the system of issuing bank-notes in France, and the system we have in England?

The precious metals are the basis of our bank-note circulation, because Sir Robert Peel, although he allowed a certain fixed amount of bank-notes to be issued against securities other than that of the precious metals, arranged that all bank-notes over this amount are to be issued against the precious metals only.

The Bank of France considers good mercantile bills approved of by the Bank a sufficient basis for the issue of bank-notes. M. Coullet, a very high authority, writes thus: " une circulation de billets de banque, basée uniquement sur les effets de commerce d'un pays, n'offre aucun péril et présente les plus grands avantàges au commerce et au public tout entier."

Now, however good the bills may be against which the bank-notes are issued, the simple principle of making the precious metals the basis of the bank-note circulation is broken through in France; and once we come to making property of other kinds a joint basis with gold and silver, we do in reality introduce a new measure of value, and we must depend on the discretion of the issuers of such bank-notes as to how far they may choose to go on their perilous path.

The Bank of France has been managed with great skill during the reign of Napoleon III, and the extraordinary financial solidity which it displays since the war of 1870—71 ought to make all Frenchmen proud of such an institution.

Mr. Ernest Seyd makes the following statement in the "Times:" "from the 13th August, 1870, when the law of prolongation came into force up to the present time (6th July 1872), on a grand total of 868,000,000 francs, or £34,700,000 of suspended or prolonged bills, there is a loss of £100,000, or scarcely one third of 1 per cent. Englishmen are proud of their financial solidity, but if such disasters, as happened to France, were to fall upon this happy land, and if the "bill cases" of our banks and the Bank of England were suddenly turned into "suspended paper," I question whether one third of 1 per cent would cover the ultimate total loss."

151. What proportion did the metallic circulation of France bear to that of the United Kingdom before the war of 1870?

Mr. Ernest Seyd assumes that the total amount of precious metals in use as money in France ranged between 330 and 350 million pounds sterling in June, 1870, and that after payment of the whole indemnity, a sum of from 125 to 150 million pounds sterling will still remain in France.

Mr. Seyd estimates the metallic circulation (including uncoined bullion) in England at 113 million pounds sterling, so that France, after paying 200 millions, will have more metallic money for purposes of exchange than England finds it necessary to keep. (London Banking, Pages 3 and 4.)

152. What may we learn by observing the great resources which France proves herself to possess?

We may, and I hope that we shall learn the importance of sufficient reserves of the precious metals being held in the United Kingdom; our people do not hoard their money as the French do, and as we all keep as little ready money by us as possible, we carry our economy in this respect too far.

" Ready money is a most valuable thing," says Mr. Hankey in his excellent treatise on Banking; "it cannot from its very essence bear interest—everyone is therefore constantly endeavouring to make it profitable, and at the same time to make it retain its use as ready money, which is simply impossible."

153. What does a contraction of the circulation lead to?

It leads to an increase in the value of the currency and a gradual fall of prices in consequence of that increase in value; we then sell more commodities to foreign nations than we did before, for they are encouraged to buy by the fall in prices, and this serves to turn the foreign exchanges in our favour.

154. What does an expansion of the circulation lead to?

It leads to a depreciation in the value of the currency and a gradual rise of prices; we then buy more commodities than we did from foreign nations, for we want to get rid of our spare cash.

155. Why is bullion exported?

Because it happens to be the cheapest commodity for us to export at that time—we can best spare it.

156. Why is bullion imported?

Because it happens to be the cheapest commodity for foreign nations to send to us at the time it is imported; besides, when our circulation begins to get contracted, standard gold begins to increase in value at home, the rate of interest rises, and money comes to where it is most wanting and bears the highest price.

157. What is the difference between a fall in price and a fall in prices?

A fall in price is a fall in the price of some one kind of commodity, and is the result of some alteration in the proportion which the supply of that commodity in some particular market bears to the demand for it there. A fall of prices is a fall in all prices generally, and is therefore the result of some alteration in the value of gold, that being the standard of value by which we measure the value of all other commodities in Great Britain. A corresponding difference exists between a rise of price and a rise of prices.

158. Has the state of the circulation any effect on the common alterations in prices which are constantly happening?

No; for these alterations in prices are not general, but confined to particular commodities; partial rises, or falls in the prices of particular commodities, arise from some alteration in the proportion which the supply of those commodities bears to the demand for them. A permanently altered state of the circulation will affect all prices without any exception, although other circumstances may frequently prevent the effect being visible.

159. What effect has the state of the circulation on the rate of interest in Great Britain?

Any expansion or contraction of the circulation will have a temporary effect on the rate of interest, causing it to rise or fall as money is plentiful or the reverse, but this effect can only last till the circulation returns to its natural state.

160. Will a steady fall in the value of gold all over the world cause the rate of interest to fall in Great Britain?

No; for the interest payable in gold will fall in value in exactly the same proportion as the principal lent in gold; temporary fluctuations in the rate of interest will continue to take place, whatever may be the change in the value of gold, for they depend upon the particular value which money has at a particular time in Great Britain, and not upon the general value of gold all over the world.

161. Will a steady fall in the value of gold have any effect on the circulation in Great Britain?

It will have the effect of numerically increasing the amount of currency in use in that country.

CHAPTER V.

GOLD.

162. Is there any such thing in reality as a standard of value which is subject to no variation?

No, there is not. If gold is chosen as the least variable standard, and consequently the one by which we are to measure the value of all other commodities, it follows that when gold rises in value all other commodities must fall, and when gold falls in value all other commodities must rise.

163. What does the value of gold depend upon?

It depends upon the proportion which the supply of gold bears to the demand for its employment. In considering the gold question we must bear in mind that the supply of gold may go on increasing for a long time without any alteration taking place in its value, *provided that the demand for its employment increases in the same proportion.*

164. What causes have been at work to increase the demand for gold during the last ten years?

A great stimulus to every kind of industry has been given by the best medium of exchange becoming more plentiful than it was before, and the commerce of the world has been much more rapidly developed on account of that stimulus. *Fresh supplies of gold, up to a certain point, have not the effect of diminishing the value of gold, but of in-*

creasing the demand for its employment. The work of colonising new countries is carried on more rapidly, railways are constructed, packet services established, docks built—in short, every kind of capital is more readily created ; and all this activity in business absorbs the gold as it arrives.

165. Up to what point do increasing supplies of gold give this stimulus to trade and business ?

Till a marked rise in the price of all commodities but gold takes place ; once this general rise in prices takes place, gold has begun to fall in value, and the only effect which further supplies of gold on the same scale can have is to cause a further steady increase in prices.

166. What separate causes may influence the exchangeable value of any commodity in a country where gold is the standard of value?

There are two causes. One is the proportion which the supply of that commodity in any particular market bears to the demand for its employment there ; the other is the abundance or scarcity of gold itself in that market.

167. What must be the first signs of any fall in the value of gold?

A marked rise in the price of all other commodities must take place. Once Europe, for instance, is saturated with supplies of gold, she will soon begin to offer less commodities in exchange for an ounce of that metal than she formerly did, and gold will fall in value. Rises in the prices of particular commodities are often attributed to a fall in the value of gold, when they ought to be attributed to the supply of those commodities being too small in proportion to the demand for their employment. Secondly, in those countries where gold is the standard of value there will be a steady numerical increase in the amount of currency required by the wants of trade.

168. Had this general rise of prices taken place in 1865 (date of third edition of this work)?

The best authorities on the subject disputed whether it had or not. We find Professor Cairnes stating in the "Times," "that a man with a fixed money income obtains now 20 per cent. less of commodities in general than he would have obtained had the gold discoveries never taken place." On the other hand, Mr. T. Crawfurd states in the same paper, "that there certainly has been no rise of price in any commodity of which the supply has been equal to the demand. There has been no rise in the price of iron, copper, tin, lead, zinc, or in rice, sugar, tea, coffee, wool, spices, or dyestuffs. In commodities there has even been a fall of price, the result of more economic processes of production" (1865).

169. If we cannot clearly prove that a general rise in prices all over the world has taken place, what conclusion must we come to?

We must conclude that gold has not yet fallen in value, and that therefore the demand for the employment of gold has been able to keep pace with the supply of that metal.

170. Must gold ultimately fall in value if the supplies continue on the present scale?

Certainly; and the time has now arrived. No one can tell how much gold will fall in value, as this depends upon the proportion which the supplies of that metal bear to the demand for their employment. Professor Fawcett, M.P., in a most valuable paper read before the British Association, referred to the great increase in the production of gold and silver during recent years; and said that although the increased supplies of gold were for a time absorbed without any decline in its value, it could now scarcely be doubted by any attentive observer, that there had been a very considerable depreciation, and that this is indicated by a marked rise in prices, which is erroneously supposed to be due not to this cause, but to a general activity in trade. The rise in prices which has occurred since 1850 is estimated by very eminent authorities at not less than 40 or 50 per cent. This partly accounts for the fact, that the apparent augmentation in the production of wealth has not produced a greater influence upon the general well-being of the whole country. If, for instance, we discover that during the past year 10 per cent. more wealth, estimated in money, was produced than in the previous year; and if, during the same period, there has been, owing to a depreciation of value in the precious metals, a rise in prices of 5 per cent., it is obvious that one half of this supposed increase of wealth is not real but simply nominal, because the prices of all commodities have advanced 5 per cent. It is not, of course, intended to be implied that there has not been a large and real increase in the production of wealth.—"*Times*," 16*th August*, 1872.

171. How long will gold continue to be supplied?

As long as it pays the expenses of its production.

172. Will it then become necessary to alter the standard value of gold in Great Britain?

No; for as long as gold is the only standard of value, it is measured in itself; and if nothing is charged on the coinage of gold, a coined ounce of standard gold ought to be exactly equal in value to an uncoined ounce.

173. What effect will a steady fall in the value of gold have on those nations who have chosen gold as their standard of value?

Creditors with debts of old standing will lose, while their debtors will gain; any one who has a fixed income payable in money must lose; there will also be some distress amongst labourers of every kind, till their salaries have been sufficiently increased; *but all the loss we can sustain in Great Britain sinks into insignificance when compared with the gain which we must make in the diminution of the burden caused by our national debt. The British people have nothing to fear from the gold discoveries.*

174. Will those nations who have chosen silver as their standard of value gain equally with nations who have chosen gold?

They will, if either the supply of silver in the world increases exactly in the same ratio as the supply of gold, or if the demand for the employment of silver is sufficiently diminished by the introduction of gold to make silver fall in value in the same proportion as gold has fallen.

175. Why has gold been chosen as the standard of value in Great Britain in preference to silver?

Because it costs much less in carriage than silver; as coin, it is more durable and more difficult for a coiner to imitate; it is also more easily counted.

176. Mention the standards of value existing at present in different countries.

There is a single gold standard in Great Britain, Portugal, Turkey, Chili, Brazils, and Australia; Germany, Austria, Denmark, and Japan, are about to adopt this standard also.

There is a double standard (both gold and silver) in France, Belgium, Switzerland, Italy, Spain, Greece, New Granada, Ecuador, and Peru. The silver dollar is still current by law in the United States, side by side with the gold dollar, although the Mint has for some time ceased to coin silver dollar pieces. There is a single silver standard in Russia, Holland, Sweden, Norway, Mexico, Central America, India, and China.

177. Is there any inconvenience in having a double standard of value?

Yes; the dearest of the two metals, where there is a double standard, will be exported whenever there is a profit to be made on the excess of the marketable value of the one over the legal value of the other. In any country where a double standard practically exists, whenever the market price of gold falls to a certain point below its legal price in silver, there is a profit to be made by sending silver out of that country in exchange for gold; on the other hand, whenever the market price of gold rises to a certain point above its legal price in silver, there is a profit to be made by sending gold out of that country in exchange for silver.

178. What proof have we that this inconvenience has existed?

One kilogramme of gold (nine tenths pure) has been made by the French law of the 28th March, 1803, equal in value to fifteen and a half kilogrammes of silver (nine tenths fine). In the first part of the present century the average marketable value of one kilogramme of gold was above its legal value in silver, and gold almost disappeared from France, for foreign merchants paid their debts to French merchants in silver, as they could get more silver for one kilogramme of gold in their own markets than they could get by law in the French markets. For instance, if a kilogramme of gold was worth sixteen and a half kilogrammes of silver at Hamburg, and a Hamburg merchant, who had gold, wanted to pay a debt to a French merchant, he changed his gold into silver at Hamburg, where he could get one kilogramme of silver more for it, and sent silver to France to pay his debt. French merchants, too, paid their debts to foreign merchants in gold as long as they could contrive to get a kilogramme of gold for fifteen kilogrammes and a half of silver at home, when its marketable price abroad was sixteen and a half. During the last ten years the average marketable value of one kilogramme of gold has been generally below its legal value in silver, and silver has been exported from France to a very large amount, for French merchants can get more gold for fifteen and a half kilogrammes of silver abroad than they can get by law at home, and foreign merchants will never pay their debts to French merchants in silver as long as they can get a kilogramme of gold for a less price in silver than that established by law in France. For instance, supposing the price of one kilogramme of gold to be fourteen and a half kilogrammes of silver at Hamburg, and that a Hamburg merchant, who has silver, owes a debt that can be paid by ten kilogrammes of gold to a French merchant, he buys the gold at Hamburg to send to France, for if he sent his silver to France he would have to send a hundred and fifty-five kilogrammes of silver, according to the French law, to pay his debt of ten kilogrammes of gold, whereas at Hamburg he can buy the ten kilogrammes of gold for one hundred and forty-five kilogrammes of silver, and save ten kilogrammes of silver in the transaction.

179. When one of the two precious metals is chosen as a standard of value, what becomes advisable with regard to the other in a country where gold and silver money are both used?

It becomes advisable to use the coins made of the rejected metals as tokens, that is, to put a sufficient quantity of alloy in them to prevent speculators taking advantage of any rise or fall in the marketable value of the standard metal as compared with its legal value in the rejected metal.

180. What is meant by the term " tokens "?

Any denomination of money the intrinsic value of which is below the nominal value assigned to it by law, and which is not a legal tender above certain small, fixed amounts.

181. Which of the two precious metals is to be preferred as a standard of value ?

In a country where bank-notes for very small amounts are permitted, as, for instance, the dollar notes in America, it matters little which of the two precious metals is chosen as the standard, provided there be but one standard. In a country where these small notes are not permitted, gold is to be preferred to silver as the one standard. A double standard cannot be resorted to with advantage in any case.

182. Why is gold to be preferred to silver as a standard of value in countries where small bank-notes are not permitted?

Because it is found more convenient to make use of silver tokens than gold tokens. Silver tokens are the best suited to make up amounts, and form but a very inconsiderable part of the circulation in countries where gold is the standard; whereas, if gold tokens were employed in countries where silver was the standard, they would enter largely into the circulation, and be made use of to pay heavy amounts, being more convenient than silver money; the only way they could be kept at the nominal value assigned to them by law, would be to make them always payable in silver on demand, and this would necessitate the expense of keeping large reserves of silver to enable the Government to keep faith with the people. Bank-notes would be to be preferred in most respects; they would cost much less, and be much harder to imitate, if proper precautions against forgery were taken; gold tokens are consequently never made use of, and in countries where silver is the standard, gold coins are allowed to circulate at their marketable value as so much metal; but this system has its inconveniences. Governments, where it is in force, must declare every six months, or oftener, at what price in silver they take gold coin in payment of the taxes. If gold falls below its declared value in silver, they lose on all the gold coin they hold in the public treasury; if it rises above its declared value in silver they gain, but their losses generally exceed their gains, as speculators are always on the watch to profit by any alteration in the value of gold; there is an opening given for such speculations to their own officers, and the public, too, always pay in whichever coin costs them least. The uncertainty as to the precise value of the gold coin in circulation leads to higgling, delay, and vexation in every-day business transactions. Gold, therefore, is to be preferred as the standard of value in countries where small bank-notes are not permitted by law.

G

183. Has gold been falling in value when compared with silver?

The diminution in the value of gold as compared with silver is hardly perceptible, nor is it likely to be permanent. The two precious metals will always fluctuate a little in their respective values from accidental circumstances; for instance, the late war in China tended to enhance temporarily the value of silver; but if gold begins to fall decidedly in value, it will drag silver down along with it. There will be two causes at work to prevent silver rising in price along with all the other commodities which are affected by the fall in the value of gold, namely, an increased activity in the silver mines of South America, and a diminished demand for the employment of silver as money. If the supply from the silver mines remained stationary, and gold was to fall in value, silver would rise in price, unless the introduction of extra supplies of gold in some way diminished the demand for the employment of silver.

184. Is there any reason for thinking that the supply of silver will increase for the future?

Yes; the principal mines on which the world depends for its supply of silver are in South America; they are of vast extent, and pronounced by scientific men to be inexhaustible; but the supply of silver from them till lately was limited by the dearness of mercury, which is used to extract the silver from the stone in which it is found by amalgamation. In 1850 a mine of mercury, called the New Almaden, was discovered in California, and other large mines have since been discovered in the Rocky Mountains, so that mercury has fallen greatly in price. This, along with improved machinery and increased facility of transport, will have the effect of steadily increasing the supply of silver from South America.

185. Is there any reason for thinking that, as gold falls in value, the demand for the employment of silver will diminish?

Yes. It must be remembered that a continued expansion of the circulation of any particular country will produce a depreciation in the value of the currency of that country, whether silver or gold be the standard of value chosen there. Gold coin is not excluded from the most of those countries where silver is the standard of value, but can circulate at its market price as so much metal; and once it is admitted to form a part of the currency, it helps to expand the circulation just as much as it would do in a country where gold was the standard. In a country where silver is the standard, gold coin must meet at first with great obstacles, from its not having a fixed value, before it can enter largely into the circulation; but as prices rise, the inconvenience of having gold coin in circulation without a fixed value will be so severely felt, that the Government will be at last obliged to fix its value, as lately happened in Belgium with regard to French Napoleons; and gold coin will not only form a considerable part of the currency, but will soon begin to take the place of silver, from its superior convenience as money. Heavy payments, too, will be made more frequently in gold bullion, and the effect will be to diminish the demand for the employment of silver.

186. If any particular country where silver was the standard, in order to preserve its currency from depreciation in value, was to refuse altogether to receive gold in payment, what would happen?

Silver would be attracted, after a little time had elapsed, from those countries where gold was the standard, or was admitted to form a part of the currency, and where, consequently, the currency was depreciated in value, to this particular country where the currency was not depreciated in value; and this influx of silver would continue till this last currency became equally depreciated in value with the others, for every one would bring their silver to the market where they could get most for it.

187. To what conclusion, then, must we come?

That, as gold falls in value, there will not only be a general rise of prices in countries where gold is the standard, but there will also be a general rise of prices in countries where silver is the standard, or, in other words, that gold in its fall will drag down silver along with it. Dr. Adam Smith alludes to the peculiar connexion existing between gold and silver in the second volume of the 'Wealth of Nations,' where he says—"Gold and silver are to be bought for a certain price, like all other commodities; and as they are the price of all other commodities, so all other commodities are the price of those metals." This fall in the value of silver, being exactly proportionate to the fall in the value of gold, will be imperceptible as far as the two precious metals are concerned; it can only be ascertained by a rise in the price of all other commodities.

188. Is there any standard of value to be found which is subject to no variation?

No, there is not; gold and silver have been chosen as standards of value, or measures by which we are to measure the value of all other commodities, not because they do not vary in value, but because they vary less in value than any other commodities do.

189. How do we find out the quantity and quality of the metal in a piece of gold?

The quantity of metal is found by weighing the piece by Troy weight—

    24 grains = 1 dwt.
    20 dwts. = 1 oz.
    12 ounces = 1 lb. Troy.

The quality of the metal is proved by a process called assaying, and is now reported by British assayers, either on the French principle of representing fine gold without alloy by 1,000 parts, called millièmes, and deducting the parts forming the alloy, or on the old method of calculation in different grades of quality called carats. By this system fine gold is gold as pure as it can be obtained, and is called 24 carat gold, while 22 carat gold, or gold of which $\frac{2}{24}$ consist of alloy, is called standard gold.

Grades of quality expressed in carats—

1 carat = 240 grains.
1 carat grain = 60 ,, or ¼ carat.
¼ ,, ,, = 15 ,, or $\frac{1}{16}$,,
⅛ ,, ,, = 7½ ,, or $\frac{1}{32}$,,

There are, therefore, 24 carats in the pound Troy, because 24 × 240 = 5,760 grains. The comparative difference in the quality of any gold, with regard to the standard metal, is called its betterness or worseness.

We give extracts from tables constructed for the purpose of valuing gold to explain each system of assaying, premising that on the carat system assayers do not go below the eighth of a carat grain, or 7½ grains Troy, so that any fraction less than an eighth is left out in their report. (See pages 87, 88.)

As the Bank price of 22 carat or 916·666 gold is £3 17s. 9d. an ounce, while the Mint price is £3 17s. 10¼d., the Bank price of fine gold, 24 carats or 1,000 fine, amounts only to £4 4s. 9·8182d. an ounce, while the Mint price is £4 4s. 11·4545d. (See the Gold Valuing Tables in Mitchell's 'Manual of Assaying,' Appendix, page 20).

"In England the 'Trade' reports of the fineness of gold bullion are made to the minimum of ⅛th of a carat grain, or 1·3 per mille, while those of the Bank of England have recently been altered to the minimum of one-third of a millième. The

THE SCIENCE OF EXCHANGES. 87

GOLD BY THE CARAT SYSTEM.

Fine Gold Per Ounce.			Carat Gold Per Ounce.			Money Value. Per Ounce.			
Oz.	Dwts.	Grains.	Crts.	Grs.	8ths.				
1	0	0·000	24	0	0	£4	4	11·4545	Better.
	19	23·375	23	3	7	4	4	10·1271	,,
	19	22·750	23	3	6	4	4	8·7997	,,
	19	22·125	23	3	5	4	4	7·4723	,,
	19	21·500	23	3	4	4	4	6·1448	,,
	19	20·875	23	3	3	4	4	4·8174	,,

And so on to

English 18 8·000 | 22 0 0 | 3 17 10·5000 Standard
         18 7·375 | 21 3 7 | 3 17 8·1725 Worse

And so on to

French   17 23·875 | 21 2 3 | 3 16 4·2436 Standard

And so on till we arrive at one-eighth of a carat grain, the lowest point we can reach.

0 0 1·250 | 0 0 2 | 0 0 2·6548 Worse
0 0 0·625 | 0 0 1 | 0 0 1·3274 ,,

Bringing the fine gold down to 0·625 grain per ounce.

GOLD BY THOUSANDTH PARTS.

	Fine Gold.	Alloy.	Money Value. Per ounce.	
	1000	·000	£4 4 11·4545	
	999	·001	4 4 10·4350	
	998	·002	4 4 9·4156	
	997	·003	4 4 8·3961	
	996	·004	4 4 7·3767	
	And so on	to		
	920	·080	3 18 1·8981	
	919	·081	3 18 0·8787	
	918	·082	3 17 11·8592	
	917	·083	3 17 10·8398	
English	916·666	·083·333	3 17 10·5000	Standard.
	916	·084	3 17 9·8203	
	915	·085	3 17 8·8009	
	And so on	to		
French	900	·100	3 16 5·5090	,,
	And so on part of	till we reach fine gold.	one thousandth	
	1	·999	0 0 1·0194	

Mint assay reports are, however, in all cases made to the tenth of a millième, or as far as it is possible to go with any degree of accuracy; and the fullest advantage, therefore, is given to the public in its dealings with the State, so far as regards the standard of gold bullion brought in for coinage. The somewhat antiquated method of calculation by 'betterness' and 'worseness' is still maintained in the Mint, and it is a question whether it should be abolished in favour of the decimal system of report, which has many advantages to recommend it. The present British standard of 916·6 per mille offers many obstacles to such a change. With the decimal standard in force in most countries of Europe, the method of decimal calculation presents no difficulties in ascertaining the fineness of bullion, and can be used with great advantage; but in this country, where the standard is 916·666 (a recurring decimal), such a method would be found to cause much unnecessary labour, and would materially increase the risk of errors of calculation."—Report for 1870 by the Deputy-Master of the Mint, page 29.

There are five standards of gold for manufactured articles—

22 carat, worth £3 17 10½ an ounce.
18 „ „ 3 3 8½ „ „
15 „ „ 2 13 1 „ „
12 „ „ 2 2 5½ „ „
9 „ „ 1 11 10 „ „

and the duty payable on gold plate or ornaments

assayed and marked by the Goldsmiths' Company, is 17s. an ounce, while the duty payable on silver plate or ornaments is 1s. 6d. an ounce; but as there is always a slight reduction of weight during the finishing processes, which follow the assaying and marking; the Government takes off one-sixth of the duty, leaving it at about 14s. 2d. for gold, and 1s. 3d. for silver.

I make the following extract from the ' English Cyclopædia' article Plate; "The assayers of the Goldsmiths' Company are not allowed to know from whose manufactured goods the particles for assay have been scraped off; each packet of particles is opened, assayed, and reported on with perfect fairness. If any manufacturer has erred frequently by sending in goods below standard, the officer who receives the goods may direct any subsequent specimens from him to be tested with additional severity. Out of London, the towns in which the largest amount of gold and silver wares is manufactured are Birmingham and Sheffield; halls, or assay offices, are established there, with duties nearly like those of Goldsmiths' Hall; and at these places a curious operation is conducted relating to 'diets' of gold and silver. An assayer scrapes eight grains from every pound troy of plate manufactured, and divides it into two portions, one of which is at once assayed; the other is placed in a receptacle called the diet-box, which at the end of the year contains specimens from all the articles manufactured.

This diet-box is sent up once a year to London, where the assayer of the Mint assays an average portion of all the fragments which it contains; if this average reaches the proper standard, the assayer at Birmingham, or Sheffield, receives a certificate; if below the proper standard, he is fined. There are a few other assay offices in the United Kingdom, each having control over the gold and silver plate made within a certain district. The Goldsmiths' Company have often petitioned for the abolition of these country offices, on the plea that the assays are more scrupulously made by the Company's servants than by those of the local halls; but it is difficult to get rid of the influence of self-interest in connection with any such plea."

The duty levied on the marking of gold is too high; and the generality of people, who would be glad to pay a more moderate duty for the guarantee of quality thus afforded, show that they think so, because they buy their gold ornaments unmarked, and trust to the tradesman who supplies the goods; yet not one of them would think of buying silver things without the hall-mark stamped on them.

There is something wrong here, and though I do not wish to interfere with the free use of gold in all its qualities as an article of commerce, I think that the public should have the option of paying a moderate duty for getting gold plate or ornaments marked at Goldsmiths' Hall. I propose a duty of fourpence per carat, no gold below fifteen carats to be marked, and no article of less than a quarter of an ounce in weight to be subject to the duty. This would amount to a duty of 5s. an ounce on the lowest quality of gold to be marked, and 8s. an ounce on the highest, or pure gold.

Once the public knew that they could get their gold ornaments marked for a moderate charge, they would soon become just as particular about the hall-mark on their gold bracelets as they now are about the hall-mark on their silver spoons.

190. How do you find the quantity of standard gold in a given quantity of gold, either better or worse than standard gold, as it may be reported?

Multiply the given quantity of gold by the number of carats in the report. Take parts for the grains, quarter-grains, or eighths of grains (should there be fractions of a carat in the report), add these parts to the sum previously obtained by multiplication. Divide the sum total by 22 (in practice it is easier to divide first by 2 and then by 11), the quotient will be the betterness or worseness of the gold, according to the report; add to it the given quantity of gold if it be a betterness, and subtract it therefrom if it be a worseness, and the remainder will be the quantity of standard gold.

191. What is the quantity of standard metal contained in 21 lbs. 10 ozs. 18 dwts. 12 grs. of gold reported worse 1 carat $3\frac{1}{4}$ grains?

lbs.	ozs.	dwts.	grs.	
21	10	18	12	for one carat.
10	11	9	6	for two grains.
5	5	14	15	for one grain.
1	4	8	16	for a quarter-grain.
39	8	11	1	

We have here divided what we put down for one carat by two, in order to find what to put down for two grains (or half a carat); we have repeated the same process to find what to put down for one grain, and we have divided what we put down for one grain by four, in order to find what to put down for a quarter-grain. The sum total is 39 lbs. 8 ozs. 11 dwts. 1 gr.

We now divide this first by 2, and then by 11:

	lbs.	ozs.	dwts.	grs.
2 |	39	8	11	1
11 |	19	10	5	12
	1	9	13	5

And this gives 1 lb. 9 ozs. 13 dwts. 5 grs. as the worseness of the gold given. Subtract this from the original quantity given:

lbs.	ozs.	dwts.	grs.
21	10	18	12
1	9	13	5
20	1	5	7

This remainder is the quantity of standard gold required.

192. How do you obtain the quantity of fine gold contained in a certain amount of standard gold?

By subtracting one-twelfth from the quantity of standard metal, a twelfth part, or $\frac{2}{24}$, being the difference between fine gold and standard gold.

193. How is gold valued which is either better or worse than standard gold?

Its quantity in standard gold is found, and then it is valued at the standard price per ounce.

194. How is the standard weight of a given quantity of silver found?

Fine silver is silver $\frac{222}{240}$ of which are pure silver, or in every 12 ozs. of fine silver there must be 11 ozs. 2 dwts. of pure silver. The ounce used is the ounce troy of 20 dwts., and the pennyweight is divided into half-pennyweights. Multiply the given quantity of silver by the number of pennyweights in the report; take a part for the half-pennyweight (if there be one in the report), add this to the sum obtained by multiplication, and divide the sum total by 222; the quotient will be the betterness or worseness, which you can then add or subtract, according to the report, from the given quantity of silver. The standard weight of silver is generally calculated to one pennyweight.

195. What is the quantity of standard metal contained in 27 lbs. 10 ozs. 10 dwts. of silver reported worse 1 oz. 17 dwts.?

Multiply 27 lbs. 10 ozs. 10 dwts. by 37, the number of pennyweights in the report. This gives 1031 lbs. 4 ozs. 10 dwts. Divide this by 222:

$$222 \overline{\smash{\big)}\,1031 \text{ lbs. } 4 \text{ ozs. } 10 \text{ dwts.}}$$
$$4 \text{ lbs. } 7 \text{ ozs. } 15 \text{ dwts.}$$

4 lbs. 7 ozs. 15 dwts. is the worseness. Subtract this from 27 lbs. 10 ozs. 10 dwts., the given quantity of silver, and the remainder is the standard weight required—23 lbs. 2 ozs. 15 dwts.

196. How is the quantity of pure silver contained in a given quantity of silver found?

Find the standard weight, multiply this by 37, and divide by 40, for standard silver is $\frac{222}{240}$ or $\frac{37}{40}$ fine.

197. How is the value of silver to be ascertained?

There is no fixed value for silver; the price fluctuates according to the state of the markets. Silver coin is generally reckoned at five shillings per ounce, and bar silver from South America at from 60d. to 62d. per ounce. The Sysee silver of China is the purest which has come to Great Britain.

198. Give an account of the present coinage in Great Britain.

Since 1816 gold has been chosen as the standard of value in Great Britain, and much care is taken to secure a proper gold coinage.

Any one may bring gold to the Mint to be coined; the Mint assayer then assays it, and if the importer is satisfied with his report, the bullion is passed into work. The Mint is bound to return in gold coin the exact weight of the standard metal which it has received; no charge for coinage being made to the importer.

The Bank of England continues to be the only importer of gold bullion for coinage purposes; private individuals do not avail themselves of this privilege, with the exception of Colonel Tomline, M.P , who sent not long ago an ingot of standard gold to the Mint to be converted into a hundred sovereigns.

The Bank buys gold with its own bank-notes, but will buy none which has not been previously assayed at the expense of the seller, by one of the six assayers employed by itself. It gives £3 17s. 9d. per ounce for all the standard gold brought to it for sale, whatever be the quantity, reserving $1\frac{1}{2}$d. on each ounce, £3 17s. $10\frac{1}{2}$d. being the exact legal price at the Mint.

Previous to purchasing, says Mr. Laing, the Bank requires bullion to be melted by one of certain firms, at present four in number. These melters run gold into ingots, or "Bank-bars," uniform in size, each containing 200 ounces. Branded with distinctive marks and figures, the bars are deposited at the Bank Bullion Office; and if the treble assay, by one of the Bank's assayers, shows them to contain not less than 21 out of 24 parts of fine gold, they are purchased. The Bank does not buy metal under 21 carats. ('Theory of Business,' page 63, second edition).

What is called the Mint price is merely a declaration of the weight of metal of a fixed purity, which the law requires to be in our gold coin; it ought to be called the Mint division, rather than the Mint price. For instance, 40 pounds troy of 22 carat gold are divided into 1,869 stamped ingots of a certain shape which we call sovereigns; whatever be the value of the metal, whether it rises or whether it falls, the number of sovereigns into which this particular quantity of standard gold is divided will remain the same, and it makes no difference whether the Royal Mint, which performs the division, is at Sydney or in London. The Bank assayer used to report to an eighth of a carat grain, or $\frac{1}{768}$ part exact, because 768 is just contained $7\frac{1}{2}$ times in 5,760 grains, equal to 24 carats or one pound troy. The importer on this system ran the risk of losing £1 in every £768, and on an average actually did lose £1 in every £1,536.

The Bank assayer at present reports to one-third of a millième, giving a risk of £1 in £3,000 or taking the actual average loss, the importer loses but £1 in £6,000. This is a great improvement; and we shall presently state its effects. (See 208.)

We give the following extract from the Coinage Act, 1870, ch. 10, sec. 8, "where any person brings to the Mint any gold bullion, such bullion shall be assayed and coined, and delivered out to such person without any charge for such assay or coining, or for waste in coinage; provided that (1) if the fineness of the whole of the bullion so brought to the Mint is such that it cannot be brought to the standard fineness under this Act of the coin to be coined thereout without refining some portion of it, the Master of the Mint may refuse to receive, assay, or coin such bullion; (2) where the bullion so brought to the Mint is finer than the standard fineness under this Act of the coin to be coined thereout, there shall be delivered to the person bringing the same such additional amount of coin as is proportionate to such superior fineness. No undue preference shall be shown to any person under this section, and every person shall have priority according to the time at which he brought such bullion to the Mint."

The working of the Mint is divided into journeys; each journey consists of 15 pounds troy weight (£701) of coined gold, or 60 pounds troy weight (£198) of coined silver.

The reason why private individuals do not have gold coined on their own account, appears to be that the loss by the delay in the coinage at the Mint exceeds the loss on selling to the Bank of England.

There is an admixture of one-twelfth of copper alloy in the English gold coinage.

A sovereign contains $113 \frac{1}{623}$ grains of fine gold; it weighs $123 \frac{274}{623}$ grains full weight, but a deficiency of $\frac{1}{2}$ grain $\frac{171}{623}$ is tolerated in circulation; that is if a sovereign does not weigh $122\frac{1}{2}$ grains, it is considered light. A half-sovereign is considered light if it weighs less than 61·125 grains. Light sovereigns will, however, continue some time in circulation, till they are remarked upon as not being of full weight, when they cease to be legal money, and the last holder must get rid of them at a loss to himself.

When sovereigns find their way to the different London banks, being deposited by customers, such as railway companies, houses of business, &c., they are sent in bags to the Bank of England, where machinery is provided to weigh each separate sovereign; those of full weight are cast into one box, and those that are light into another; the banks, which have sent the bags, are charged with the loss on these last, and they are at once transferred to other machinery which cuts them in two. In the three years ending 31st March, 1872, gold coins of the nominal value of £1,975,716 were cut by the Bank. The loss sustained by the owners of the coins amounted to £25,415, a seventy-eighth part of the nominal value. The Bank used to give but £3 17s. 6½d. an ounce for light gold coin, and the loss thus entailed on the last holder acted as a positive inducement to bankers, houses of business, &c., to keep light sovereigns in circulation instead of trying to withdraw them as they ought to do. The Bank now gives £3 17s. 9d. an ounce for light gold coin, ranking it as standard gold which the Mint has solemnly pronounced it to be. The Mint gives £3 17s. 10½d. an ounce for light gold coin, and will pay the value of the importation either in new gold coin or by cheque on the Bank of England; no parcels of a nominal value of less than £100 will be received.

Bankers and houses of business would soon avail themselves of this excellent arrangement, and send in their parcels, were it not for the sixth regulation in force at the Mint, "the Master of the Mint cannot undertake to return light gold coin recoined, or to make any payment on account of its delivery for re-coinage until a period of fourteen days shall have elapsed from the date of its importation," and this period may be subject to extension because the maxim of "first come first served" is followed at the Mint, or because the amount of gold in that establishment is not sufficient to justify the Master in resuming coinage operations.

The Bank of England is now the sole importer of light gold coin at the Mint, and for exactly the same reason that it is the only importer of gold bullion for coinage purposes.

The public prefer to get £3 17s. 9d. an ounce for the light gold coin or bullion which they have to sell, and to be paid on delivery, than to get £3 17s. 10½d. an ounce with the delay taking place at the Mint.

The Mint ought to be obliged by law to take light gold coin of the realm by weight at £3 17s. 10½d. an ounce, and to pay for it on delivery by cheque on the Bank of England, and the expense of re-coining such light gold should be borne by the State.

"The average gold coinage for ten years, up to and including 1866, slightly exceeded £5,000,000 each year. The year is generally speaking nearly equally divided between gold and the other metals. The silver and copper coins more than repay their cost by the seignorage on these coins. The average cost of coining a sovereign would be approximately obtained by dividing the Mint expenditure for six months over 5,000,000 sovereigns. The expenditure being about £15,000, the outlay on each sovereign would appear to be 0·72 penny or ¾d. each" (Evidence of R. A. Hill, Esq., Blue Book, page 44).

As we use both silver and copper money, though we have chosen gold as our sole standard of value, we are obliged to put a sufficient quantity of alloy in our silver and copper coin to prevent speculators taking advantage of fluctuations in the relative values of the three metals; for instance, we make a bronze penny do the duty of an intrinsic copper penny, though it is intrinsically worth about a quarter of such a coin, because we have found by experience, that, if we had a copper coinage like our gold coinage, speculators would take advantage of any rise in the marketable value of copper, as compared with its legal value as coin when measured in gold money, and melt all they could get of it for sale as metal. Our present pence are a legal tender in payments to the amount of 1s. only; half-pence and farthings to the amount of 6d. only. The material of this coin is a bronze mixture composed in 100 parts by weight of 95 copper, 4 tin, and 1 zinc, the same as in the bronze coinage of France.

The same reasoning applies to our silver coin

also; we could not preserve it from the manipulations of speculators unless we put it into circulation at a lower intrinsic value than the nominal value assigned to it by law.

Silver coin in Great Britain is not a legal tender above 40s.; silver, of which 222 parts are pure silver and 18 parts alloy, is standard silver, and a pound troy of such silver is coined into 66s.

"The average market price at which standard silver bullion has been purchased by the Mint during the year 1871 is 5s. $0\frac{7}{8}$d. per ounce, as against 5s. $0\frac{1}{2}$d. per ounce in 1870, so that the rate at which silver coin is issued by the Mint being 5s. 6d. an ounce, the seignorage accruing to the State has been at the rate of 5 $\frac{9}{10}$d. per ounce or nearly $9\frac{1}{4}$ per cent." (Second Report, page 7).

This percentage is retained by Government to defray the expense of the coinage and to keep it in repair, for Government are bound to give new silver coins in exchange for any which may have received injury in circulation, provided that injury has not occasioned the complete defacement of the coin, that is to say, as long as enough of the superscription remains to warrant to what country the coin belongs. The Bank of England notifies to Government when silver coin is wanting, and they employ brokers to buy silver bullion in the open market; it is then coined, and sent to the Bank, and as it is absorbed in the circulation, Government is gradually repaid by the Bank. If silver was to rise much in value as compared with gold, Government would be compelled to put less silver in its coins than it does now; for if it continued to issue silver coin at 5s. 6d. an ounce, or 66s. to the pound troy, and silver were to rise to 6s. an ounce in the open market, speculators would soon begin to melt the coin, or to export it as metal. On the other hand, if silver was to fall in value as compared with gold, it would be a wise policy on the part of Government to put more silver in its coins than it does now. The only remedy for an excess of silver money lies in the value which it actually possesses as silver bullion, and we ought to be very careful to keep our silver coinage at that precise intrinsic value below its nominal value by law, which will prevent its being melted by speculators, but not to venture below this limit. A time might come when we wanted every pound of our silver coinage to keep our flag aloft.

We shall now give a short theory of the coinage. Standard coins, whether they be of gold or silver, are but the divisions of standard metal which each nation has chosen to make, in order to suit its own convenience. These divisions may of course be altered by law, but such alteration has no effect on the actual value of the standard metal.

For instance, we in England now divide 40 lbs. troy of standard gold into 1,869 ingots called sovereigns; we might by legislation divide the same quantity of metal into 2,000 ingots and call them sovereigns still, but this would have no effect on the value of our standard metal; 40 lbs. troy of 22 carat gold would have exactly the same purchasing power all over the world which it had before, but our new sovereigns would have less gold in them, and would therefore be able to purchase less of all other commodities. No law can make an ingot containing 112 grains of 24 carat gold permanently equal in value with one containing 113 grains of the same metal.

We give a value by legislation to our silver and bronze coins which they do not possess as ingots of metal, but they keep this arbitrary value for two reasons—first, because they are a legal tender for small amounts only; secondly, because they are not put into circulation unless there is an evident demand for them, so that they are always exchangeable for our standard coin at the rate fixed by law, and at the convenience of the holder. Were these limitations removed, our silver and bronze coins would soon lose their arbitrary value.

199. Was there any justice in the complaints made by Colonel Tomline, M.P., as to his not having the power to get silver bullion coined at the Mint?

There was this much justice; the silver coinage had not been supplied in sufficient quantity, and there was some inconvenience occasioned to the public in consequence.

Silver bullion, with our present monetary system, cannot be coined on demand for any one who chooses to bring it to the Mint, because the profit on the transaction is large, and it would soon lead to great accumulations of silver coin in the hands of bankers, railway companies, &c., which, as last holders, they could not get rid of without loss, silver coin not being a legal tender in amounts above 40s.

The inconvenience, however, is of a temporary nature, and the best means of preventing its recurrence would be for the Bank of England to extend its present system of obtaining information as to existing supplies of silver coin to the British Colonies as well as to the United Kingdom, and to make use of its power of issuing banknotes against silver bullion to a limited extent, by purchasing it whenever it is to be bought at a low price, and by then holding it under an indemnity from Government as a reserve to supply the Mint with working material from, so that this last establishment may be kept regularly at work with as little loss as possible. There is no likelihood at present of a want of work at the Mint, because the depreciation in the value of gold will necessitate a large numerical increase in the amount of standard coin and subsidiary currency, which we want for the United Kingdom and the Colonies; besides, as the "Economist" says, our gold coinage is in such repute, that we manufacture it to a large extent for export.

As to the second part of Colonel Tomline's complaints, that the coinage of gold is free, while a large seignorage is charged on the manufacture of silver coin, which poor people use so largely, and that this is unjust because it favours the rich, the gallant colonel has no ground to stand on.

The standard of value in the United Kingdom is 22 carat gold, and each silver coin which we use represents a certain portion of the standard gold ingot which we call a sovereign. A shilling, for instance, represents the twentieth part of a sovereign; we give it this power of representation quite irrespective of the metal contained in it, and we call it very properly a token.

"It is quite true," says Mr. Hubbard, "that silver rather than gold is the medium through which the wages of the labouring classes are paid; but to show that the labouring classes are injured by the Mint regulations, it must be demonstrated that the shilling they now receive commands a smaller quantity of the necessaries of life than would a shilling coined as an integral measure of value. The shilling now circulating derives its purchasing power, not from the silver it contains, but from its being by law a twentieth part of a pound—the golden standard. All prices, wholesale or retail, whether of a bullock or a beef-steak, of a quarter of wheat or a loaf of bread, are computed upon a gold valuation. The artisan's shilling is intrinsically the twenty-second part of a pound, his penny but the four hundred and eightieth part of a pound; but how do these facts affect his interest, if he can always with twenty shillings or two hundred and forty pence secure the value of a pound?

If a silver standard of value were substituted for or made concurrent with our present gold standard, the great object to be studied would be so to adjust the new standard in its relation to the old one, as to leave the prices of commodities practically unaffected."—" *The Times,*" 18*th* November, 1871.

We cannot keep gold standard coins and silver standard coins in circulation side by side in any one country for a continuance, because the value of one metal when measured in the other is sure to fluctuate; and, as all standard coins are but so many stamped ingots of standard metal, people will always select the metal which costs them least when they have a payment to make; so that if more can be got for a given quantity of one metal in foreign countries than can be got for its legal equivalent in the other metal, supposing them to be circulating as stamped ingots side by side, the metal for which the most can be got will be exported, because of the two it is the cheapest mode of payment.

For instance, if eleven rupees' worth of goods can be got in England for a sovereign, which is circulating side by side with rupees as ten in British India, the sovereign will be exported, and the rupees will be kept in India.

On the other hand, if one sovereign's worth of goods can be got in England for ten rupees, when rupees are circulating at the rate of eleven to one sovereign side by side with sovereigns in India, the rupees will be exported, and the sovereign will be kept in that country. In any free country what is worth more abroad than it is at home, is certain to be exported, whether it be coin or cotton.

There ought to be one standard of value in any one country, and but one; and the standard coins should circulate as stamped ingots of the standard metal, no charge being made for the actual cost of their manufacture.

There are some very scientific economists who do not agree with this doctrine; and the proof of the consideration given to their opinion lies in the fact that in the last monetary convention (23rd December, 1865), to which France, Italy, Switzerland, and Belgium were parties, five-franc silver pieces are still allowed to circulate as standard coin, 900 fine, side by side with gold standard coin of the same fineness; so that debtors in these four countries may select either five-franc silver pieces or gold coins to make their payments in, and will of course select the metal which costs them least.

By the same convention two-franc pieces, and all silver coins under two francs, have been constituted tokens, and each of the contracting nations is allowed to coin only a "pro rata" amount of such tokens.

Spain, though not bound by the Convention, appears to be inclined to act upon the principle carried out in it, because she has selected the Peseta as her monetary unit, equal to 100 centimes; and is about to coin standard gold coins of 100, 50, 20, 10, and 5 pesetas, 900 fine; also 5 peseta silver pieces of the same fineness, while her subsidiary coins are to be tokens.

Austria is about to introduce a gold standard

coinage of eight florin and four florin pieces, 900 fine, equal in value to the twenty-franc piece.

Germany has selected the metal gold as a standard, and her gold coins called the 20-mark and the 10-mark piece, are to be struck at 900 fine; the mark being divided into 10 groschen, and the groschen into 10 pfennigs. This mark is just equal to 11¾d. of our money, so that the 20-mark piece will be worth 19s. 7d.

We shall, therefore, have three leading gold coins in use throughout Europe; the 20-mark piece for Germany, the 20-franc piece for France, Italy, Switzerland, Belgium, Spain, and Austria, and the sovereign for the United Kingdom. We shall also have two different qualities of gold used as standard metal; the British standard 916·666, and the Foreign standard, 900 fine, adopted by all the other States of Europe, and by the United States also.

In France the proportion between the respective values of silver and gold is fixed by law to be as 1 is to 15½, or 1 kilogramme of gold to 15½ kilogrammes of silver. The expense of the standard coinages is borne by whoever brings the metal to the Mint, whether it be silver or gold; a kilogramme of silver, five franc pieces, is coined into 200 francs, at a charge of 1 franc 50 centimes, and a kilogramme of gold is coined into 155 Napoleons, at a charge of 6 francs 70 centimes; to coin, therefore, its equivalent in silver, or 15½ kilogrammes of silver, costs 23 francs 25 centimes, a difference of 16 francs 55 centimes in favour of gold coin. The French Government, in 1803, believed that it had established one sole measure of value, to wit, the franc, whereas, by permitting a double-paying medium, with a fixed proportion between the two precious metals, it practically created a double standard of value. As long as the marketable value of one kilogramme of gold, estimated in silver, continued above its legal value in silver in France, the French people used silver money, for the simple reason that it cost them less than gold; but once the marketable value of one kilogramme of gold, estimated in silver, fell below its legal value in silver in France, the French people began to use gold coin instead of silver, for the simple reason that it cost them less than silver. The dearest of the two metals (when there is a practical double standard) will always be exported. The bronze coins are tokens like the British copper money, and are not a legal tender over 50 centimes. The French mode of reporting gold and silver is to report the whole of the fine metal in millièmes, or thousandth parts of the metal under trial, and their assayers report to $\frac{1}{10,000}$ exact. A kilogramme of coined silver is valued at 200 francs, and as the proportion between it and coined gold is fixed by law at $15\frac{1}{2}$ to 1, a kilogramme of coined gold is valued at 3100 francs; if, then, we subtract the charge made for expense of coinage, the prices paid by the French Mints are 3093,30 francs per kilogramme for standard gold, and 198 francs 50 centimes for standard silver.

THE SCIENCE OF EXCHANGES. 111

To reduce a French report in millièmes to an English report in carats, we multiply the number of millièmes by 24 for the carats, and by 4 for the carat grains, remembering that we are not to go below the eighth of a carat grain, or $7\frac{1}{2}$ grains troy:

.950 millièmes.
× 24

———

3·800
19·000

———

22·800 carats
× 4

———

3·200 carat grains.

                      carats. grs.
Report ·950 fine = 22 $3\frac{1}{8}$
British standard = 22

———

Better . . . $3\frac{1}{8}$

French gold, report ·950 fine = British gold better $3\frac{1}{8}$ carat grains.

To put British gold better $3\frac{1}{8}$ carat grains into the corresponding French report we reverse the operation, and divide 3·200 carat grains by 4 to get the decimal of the carats, and then divide the carats themselves and their decimal by 24 to get the millièmes, thus :

carats. grs.

4) 3 · 200 (·800
   3 200

———

24) 22 · 800 (·950 millièmes.
    21 6

———

    1 20
    1 20

———

British gold better $3\frac{1}{8}$ carat grains = French gold, report ·950 millièmes fine.

The tariff rates for millièmes of gold and silver begin with 1000 as their highest expression in each case, better 2 carats than British standard gold, and better 18 pennyweights than British fine silver, thus:

For gold. Millièmes.	Tariff rates at French Mints. Francs per kilo.			British reports. Carats. carat grs.	
1000	. .	3437	Better	2	
999	. .	3433	56	,,	1 $3\frac{7}{8}$
998	. .	3430	12	,,	1 $3\frac{6}{8}$
997	. .	3426	68	,,	1 $3\frac{5}{8}$

And so on till we come to British standard gold:

916 . . 3148 29 Standard 0 0

And to French standard gold:

900 . . 3093 30 Worse 0 $1\frac{5}{8}$

The intrinsic value of a kilogramme of pure gold in France, without any deductions being made, is fr. 3444·44, and of a kilogramme of pure silver, fr. 222·22.

For silver Millièmes.	Tariff rates at French Mints. Francs per kilo.			British reports. Pennyweights.	
1000	. .	220	56	Better	18
999	. .	220	33	,,	$17\frac{3}{4}$
998	. .	220	11	,,	$17\frac{1}{2}$
997	. .	219	89	,,	$17\frac{1}{4}$

And so on till we come to British fine silver:

925 . . 204 01 ,, 0

And to French standard silver :

For silver Millièmes.	Tariff rates at French mints. Francs per kilo.		British reports. Pennyweights.
900	198	50	Worse 6

The prices per kilogramme in francs are found thus; we multiply the price for either French standard gold or silver by the number of millièmes the price of which we want to find out, and divide by the standard number of millièmes.

For instance, the price in francs per kilo for gold 916 fine is :—

Francs.
$$\frac{3093\cdot30 \times 916}{900} = \text{francs } 3148\cdot29$$

For silver 925 fine—
$$\frac{198\cdot50 \times 925}{900} = \text{francs } 204\cdot01$$

In doing these sums, if 5, or above it, remains after the division, it makes the rate another centime. There are regular tables of tariff rates for millièmes and British reports, constructed to shorten exchange calculations. — (See " Bullion and Foreign Exchanges," pp. 152, 153).

As I am treating of the silver coinage of the United Kingdom in this answer, I may remark that the issue of half-crowns from the Mint has been suspended since 1851, and that the public are not only deprived thereby of a most convenient coin, but that a large additional number of sixpences must be coined to do the duty of those contained in the half-crowns. It is just as easy to count by eighths of a sovereign as by tenths, and when we have a good old coin like a half-crown which we all like to see, and which can stand wear and tear well, we ought to keep it. " Videant poveres et lætentur."

CHAPTER VI.

THE FOREIGN EXCHANGES.

200. Give an instance of exchange when it is said to be favourable, when adverse, and when at par between two places in different countries.

(1) When you pay a sum of money in London which contains ten ounces of standard gold by mintage valuation, for a bill which you can have cashed in New York for a sum of money which contains an equal quantity of standard gold by mintage valuation, exchange is said to be at par between London and New York.

(2) When you pay less in London for such a bill, it is said to be at a discount, and exchange is said to be in favour of London, and against New York.

(3) When you pay more in London for such a bill, it is said to be at a premium, and exchange is said to be against London, and in favour of New York.

The true par of exchange, therefore, between any two places in different countries, is the being able to obtain a quantity of standard gold or silver, when you cash a bill in the one place, exactly equal to the quantity of standard gold or silver which you have paid for the same bill in the other; it is *not* the merely obtaining an equal nominal amount of money, for if the currency of any country is much depreciated in value from over-issues of paper money or a base coinage, so much the more of it will be required to make up the necessary amount, and this in exact proportion to the extent of the depreciation in value which has taken place.

201. What is a rate of exchange?

It is the price of the money of any one country reckoned in the money of any other country.

202. What is the course of exchange?

The fluctuations in the rate of exchange above and below what is considered to be par between any two places in different countries are called the course of exchange between those places. When two countries have one and the same standard the par of exchange between them is on a solid, fixed basis; but when one country has gold for its standard and the other has silver, the number of ounces of standard silver in the one country which are to be considered equal to an ounce of standard gold in the other must first be fixed, and the par of exchange is on a basis which must fluctuate slightly as the relative proportion in value which the precious metals maintain towards each other fluctuates.

203. Explain the course of exchange as quoted in the newspapers.

If the rate of exchange is considered with regard to any two places, it will be seen that the price given for any particular sum of money may vary, but that the particular sum of money to be bought is always the same, though more or less may be given for it. For instance, Paris may give London 25 francs 20 centimes, or 25 francs 30 centimes for a pound sterling, but whatever Paris gives, the pound sterling is fixed and invariable, though the price paid for it may vary. The place which pays a variable price in its own money for a fixed sum in the money of some other place is said to give so much of its own money, while the place to whom that variable price is paid for the fixed sum in its own money is said to receive so much of the money of the other place; thus, London gives St. Petersberg $38\frac{1}{4}$ pence for one silver rouble, but receives 6 dollars $24\frac{1}{4}$ silver groschen from Berlin for a pound sterling. (See next page.)

204. How many kinds of exchanges are there?

There are two—direct, and arbitrated. Direct exchange is the operation of exchange between any two places in different countries, without the medium of any other place, as between London and Paris. Arbitrated exchange is the operation of exchange between any two places in different countries, through the medium of some other place or places in other countries.

205. What is the object of bankers and cambists in all these operations?

It is to choose whatever mode of payment costs them least; for remittances abroad they will choose whatever mode of payment gives the greatest sum in foreign money for a given sum in the money of their own country, and for returns to their own

COURSE OF EXCHANGE.

MONEY AT	LONDON RECEIVES FROM, OR GIVES TO, *Variable Rate.*	*Fixed Rate.*
Amsterdam, 5 cents = 1 stiver, 20 stivers = 1 guilder.	Amsterdam, 12 florins $0\frac{1}{3}$ stivers ..	.. for one pound sterling.
Berlin, 30 silver groschen = 1 thaler.	Berlin, 6 dollars $24\frac{1}{4}$ silver groschen	.. for one pound sterling.
Calcutta, 12 pie = 1 anna, 16 annas = 1 rupee.	Calcutta, $22\frac{3}{8}$ pence English.. ..	.. for one rupee.
Christiania, 120 skilling = 1 specie-daler.	Christiania, 4 specie daler $58\frac{2}{3}$ skilling	.. for one pound sterling.
Constantinople, 40 paras = 1 piaster, 100 piastres = 1 medjdie.	Constantinople, 110·780 piastres ..	... for one pound sterling.
Copenhagen, 96 skilling = 1 Rigs daler.	Copenhagen, 8 Rigs daler 94 skilling	.. for one pound sterling.
Frankfort, 60 Kreuzer = 1 Zolverein florin.	Frankfort, 119 florins 10 kreuzer ..	.. for ten pound sterling.
Genoa, 100 centesimi = 1 lira Italiana.	Genoa, 25·2215 lire	.. for one pound sterling.
Hamburg, 12 pfenning = 1 schilling, 16 schilling = 1 mark.	Hamburg, 13 mark $7\frac{1}{2}$ schilling ..	.. for one pound sterling.
Japan, 100 sens = 1 yen.	Japan, $245\frac{1}{2}$ pence English	for the piece of five yens (gold 900 fine).
Lisbon, 1000 reis = 1 milreis.	Lisbon, $53\frac{1}{3}$ pence English	.. for one milreis.
Madrid, 34 maravedis = 1 real, 20 reals = 1 duro.	Madrid, $49\frac{1}{2}$ pence English	.. for one hard dollar.
New York, 100 cents = 1 dollar (the almighty dollar).	New York, 49·316 pence English ..	.. for one dollar.
Paris 100 centimes = 1 franc.	Paris, 25 francs $22\frac{1}{2}$ centimes ..	.. for one pound sterling.
Pekin, 1 tael = 10 mace = 100 candareens = 1000 cash.	Pekin, $71\frac{3}{4}$ pence English	.. for one tael.
Rio di Janeiro, 1000 reis = 1 milreis.	Rio di Janeiro 26·93 pence English	.. for one milreis.
St. Petersburg, 100 copek = 1 silver rouble.	St. Petersburg, $38\frac{1}{8}$ pence English	.. for one silver rouble.
Stockholm, 100 ore = 1 riksdaler riksmynt.	Stockholm, 17·80 riksdaler	.. for one pound sterling.
Vienna, 100 kreuzer = 1 florin.	Vienna, 10 florins $21\frac{1}{2}$ kreuzer ..	.. for one pound sterling.

The variable rates are generally only given in the published quotations; for instance, Madrid, 3 months $48\frac{3}{8}$ to $\frac{1}{2}$; Frankfort on the Maine $119\frac{3}{4}$ to 120; Oporto, ninety days $52\frac{3}{4}$ to $\frac{7}{8}$; Vienna 11·35 to 11·40, and so on.

country they will choose whatever mode gives the greatest sum in the money of their own country for a given sum in the money of any foreign country. What is bad, therefore, for remittances, will be good for returns, and *vice versâ;* for instance, the rate of 25 francs 70 centimes per £1 is a good rate for remittance to Paris from London, because you get more foreign money than usual for a given sum of your own money, but it is a bad rate for a return of money to London, because you get less of your own money than usual for a given amount of foreign money. If we want to buy lemons at Palermo, the more of them we can get for a sovereign, the better for us; but if we want to sell lemons there and obtain a sovereign for them, the fewer the lemons we have to give to get that coin, the better for us. If we substitute Italian lire for the lemons we shall see that when we want to sell sovereigns and to buy lire, the more lire we can get the better the exchange is for us; while if we want to sell lire and to buy sovereigns, the fewer lire we have to give to get the number of sovereigns required the better for us the exchange is. Hence an Englishman buying sovereigns at Palermo would like very well to obtain them at 24 lire Italiane each, while an Englishman selling sovereigns there would like to get 26 lire Italiane for each of them if it could be had. The price which the operator would like to pay, varies according to the nature of the transaction. The terms cash and bills are generally used instead of remittances and returns; thus, when 25 francs 70 centimes can be had for £1 in Paris, it is said to be better for cash in London, that is, the laying out of cash in London, and worse for bills or receiving returns from Paris, because less English money can be then obtained for a given sum of French money. The amount of a bill payable in a foreign country is usually expressed in the money of the country in which it is to be paid, and the rate for brokerage is generally a tenth per cent. in the principal places of Europe.

206. Is there any limit to the extent of the rise above or the fall below par, as the exchange between London and any other place is favorable, or the reverse?

In ordinary cases the limit is the total expense of transmitting bullion from the one place to the other; by total expense we do not mean the mere carriage of bullion, for, except in times of war, this will vary hardly at all, but we mean all charges whatever connected with the transmission of bullion, including the calculation of the rates of interest obtainable for the money represented by the bullion in the country it is sent from and in the country it is sent to. If a merchant had to pay *more* for a bill on any place than it would cost him to transmit bullion or coin to that place, he would not buy the bill, but would send the bullion or coin instead; on the other hand, a banker or cambist who holds a bill will not generally take *less* for it than it would cost him to send it to the place on which it is drawn, have it cashed there, and the proceeds sent to him in coin or bullion.

The higher the rate of interest rises at home the more it will cost to send bullion or coin abroad, as the price which could have been had for its use as money at home must be taken into consideration, and, except in the case of a loan, it will only be transmitted from one country to another when this is the cheapest mode for the debtor-country to satisfy its foreign creditors, either from its having superfluous cash at home or from its having no other mode of payment immediately available for the purpose.

"Merchants," says Jean Baptiste Say, "find out the difference of purchasing power, which the money of any one country may possess in any other country, by comparing what that money is able to procure for them there with what other kinds of merchandise would procure, if sent instead of money."

"A merchant who is considering whether he will export piastres or Malaga wine from Spain to France, compares the quantity of any particular commodity which a thousand piastres will procure for him in France, with the quantity of the same commodity which Malaga wine to the value of a thousand piastres in Spain would procure for him if exported to France.

"If, for instance, a thousand piastres when sold in France would be able to purchase a hundred pieces of cloth of Brittany, and if Malaga wine, worth a thousand piastres in Spain, and exported to France to be sold there could only purchase ninety-six pieces of the same cloth, there would be a gain of four per cent. on sending the piastres instead of the wine, supposing the same cost of transmission in each case.

"When the precious metals, from the effect of payments of money which we have to make to other countries, become scarce in our own, so as to raise their value at home by two or three per cent. only, our merchants are directly interested in importing them; but our merchants cannot import the precious metals without paying for them, without exporting some equivalent in the produce of our own country.

"It is to the last degree plain that we can pay our debts with our produce only, or, what comes to the same thing, with what we acquire by the sale of our produce.

"A country like Mexico, which pays for its purchases abroad by the export of silver bullion, is still paying its debts with the produce of its soil and its industry, for silver bullion in such a case forms a part of the produce of the country."— (Notes 24 and 25 Catechism, 2nd edition.)

207. Give an explanation of the valuing of Gold?

What we call the Mint price is a division and not a price, because gold is divided in the same way at the Royal Mints; nor has the value of the sovereign anything to say to this division, whether sovereigns rise or fall in value, whether the Mint be in London or at Sydney, 40 pounds troy of gold 22 carats fine are divided into 1,869 standard ingots or sovereigns.

However largely the production of gold in Australia might be increased, the only effect this would have would be to make it fall in value as compared with other commodities; more of it would then be sent to London; and this, if continued long enough, would make gold in London fall in value also.

Once the trading communities of the world have enough gold at a certain value to carry on their operations with, any further supplies can only have the effect of making the metal fall below that value, as compared with all other commodities. Where three sovereigns had a certain purchasing power, four will then be required, and so on.

"By Mr. Dunlop's statement," says Colonel Smith (page 51, " Remarks on a Gold Currency "), "Australian gold costs the same, if not more, on delivery in London, as it does landed in India, namely £3 18s. 4d. per standard ounce; nevertheless, the greater part of it is sent to England, and sold to the Bank for £3 17s. 9d. per ounce only."

Such a statement may appear puzzling at first, but we think it can be made quite clear by the following instance:—Let us suppose a wool merchant in Australia to have some reason for setting up business as quickly as he possibly can in London. He goes to another wool merchant at Sydney, and he says, "I know you have a good lot of wool on its way to London, I shall be glad to give you some of mine which I have here in exchange for it, and you will give me an order to enable me to take yours when it arrives in London." If wool in London is worth one-tenth more than it is in Sydney, his friend answers, "Your wool and mine are equally good, therefore I have no objection, but for every ton I give you delivered in London, you must give me a ton and 2 cwt. delivered here." Wool is worth more in London than it is in Sydney, and therefore, to get a ton of wool delivered in London, more than a ton, though the wool is equally good, must be given at Sydney. It is exactly the same with gold; it is sent from Australia because it is worth more in London, and therefore, to secure the delivery of 10 ounces of 22 carat gold in London, somewhat more than 10 ounces of the same gold must be given in Sydney, so that where gold is worth the least, the quantity of it which we have to give to secure its delivery in the countries where it is worth most, must be always greater than the actual quantity which is delivered in such countries. For instance, if 22 carat gold in London is worth one-tenth more than it is at Sydney, and, if we exclude the expenses of transmission each way, to get 10 ounces delivered in London we must give 11 ounces at Sydney. On the other hand, if we are in London, and want 10 ounces of 22 carat gold delivered in Sydney, we need only give $9\frac{1}{11}$ ounces. As the Mint price or division of gold is exactly the same in each place, our only way of expressing the nature of the transaction is by adding or subtracting from the Mint price in

London or in Sydney; we must add one-tenth of £3 17s. 10½d. to this sum itself, to represent the price of gold in London bought at Sydney, while we must subtract one-eleventh from £3 17s. 10¼d. to represent the price of gold in Sydney which is bought in London. We thus arrive at the Australian and Indian price of £3 18s. 4d.

It was this difference in the value of the same quantity of standard gold in two different countries at the same time which Mr. Goschen was thinking of when he said, "if we could get rid of all fluctuations of exchange, that would indeed be a very agreeable result. But even if the piece of twenty-five francs was made equal to the sovereign, and the two coins were absolutely identical, were two sovereigns in fact, the sovereign in France would not necessarily be equal to the sovereign in England. It would still depend upon the balance of trade; upon the demand for gold for remittance to England, or for remittance to France; you would never get rid of the fractions."

I am decidedly against the introduction of an international coinage, but I agree with Mr. Jacob A. Franklin in thinking that an international Mint code might be very useful, whereby all bars of gold held in stock, and prepared for coinage, should be of one weight, quality, length, and thickness, and have their value stamped upon them by the Mint they came from; they would prove a very convenient means of remittance, as they could be sent in sealed parcels from one Mint to another.

Our credit, both at home and abroad, depends on the quality of our gold coinage, and we ought therefore to trust to none but our own Royal Mints for its manufacture, and to none but a Pyx Jury of English Goldsmiths for the verification of its weight and fineness. " The object of the trial of the Pyx is to satisfy the public that the coins issued from the Mint are accurate, both in respect of weight and fineness, and this guarantee is furnished by means of a periodical examination made by the Freemen of the Goldsmiths' Company, under the direction of the Crown." (First Report of the Deputy-Master, page 34.)

The trial of the Pyx is to be held once in every year in which coins have been issued from the Mint.

One coin is taken from each journey weight, and the gold and silver coins carefully laid by in sealed packets. The jury then take from each packet as many coins as they think necessary for the purpose of the trial; they first weigh these coins, and then they assay them individually to ascertain their fineness.

"The jurors having assembled and their names having been called over, the Queen's Remembrancer administers the following oath to them :—' You shall well and truly, after your knowledge and discretion, make the assays of these moneys of gold and silver, and truly report if the said moneys be in weight and fineness according to the standard weights for weighing and testing the coins of the realm, and the standard trial plates of gold and silver used for determining the justness of the gold and silver coinage of the realm in the custody of the Board of Trade, and be in conformity with the first schedule of the Coinage Act, 1870. So help you God.'

Before dismissing the jury to the performance of their duties, the Queen's Remembrancer addresses a few remarks to them, specially pointing their attention to the schedule mentioned in their oath, and to the instructions generally contained in Her Majesty's Order in Council.

The following extract from the first schedule of the Coinage Act, 1870, will show at a glance the delicacy of test to be observed by the jury in weighing and assaying the various moneys submitted to their trial, and the closeness to Standard to which the officers of the Mint are compelled to work. The Imperial weight in grains only is here given :—

Coin.	Weight.	Remedy.
Sovereign	123·27447	0·20000
Half-sovereign	61·63723	0·10000
Florin	174·54545	0·72727
Shilling	87·27272	0·36363
Sixpence	48·63636	0·18181

" The standard fineness for gold coins is 11-12ths fine gold and 1-12th alloy, or millesimal fineness 916·66 ; the remedy, or margin allowed to the Master of the Mint on coinage being millesimal fineness 0·002. For silver coin the standard fineness is 37-40ths fine silver and 3-40ths alloy, or millesimal fineness 925·, the remedy being millesimal fineness 0·004.

" Before the jurors proceed to the laboratory to commence their operations, they are furnished with small portions of the Standard trial plates, in the custody of the Board of Trade, to guide them in their assay of the coins deposited in the Pyx.

" At the hour named by the jury for rendering their verdict, the Queen's Remembrancer again attends at Goldsmiths' Hall to receive it."—(From the *Times*.)

The jury also weigh the residue of the coins left in the packets in bulk to ascertain that they are within the remedy for weight.

208. What is the effect of a mintage or charge on the manufacture of standard coin?

As long as any currency continues at its usual value as compared with the other currencies of the world, a mintage, or import duty on bullion turned into coin, paid by bearers of gold to the Mint, will add to the value of the gold currency in question. It therefore acts as a bar to prevent the coin being exported, because during this particular state of the circulation, the coin is worth more in its own country than in any other. But let us suppose the currency to expand for a time, and therefore to become depreciated in value from redundancy, the coin gradually sinks to its actual intrinsic value as metal, while the depreciation is going on. It is then exported or melted until a gradual contraction of the circulation restores the currency to its usual value. The increased value, therefore, which a mintage gives, cannot be always maintained, because when the circulation of any country is contracted its coin will be worth more, and when it expands and becomes redundant, its coin will be worth less.

"It is my opinion," says Mr. Goschen, M.P., "that although at the time of coining, the coin is more valuable than the bullion from which it is made (at any rate in the opinion of the person who carries the bullion to the Mint to be coined), that increased value cannot be permanently maintained, and under certain circumstances, of not unfrequent occurrence, actually disappears." (Blue Book on The Coinage, page 124.)

M. Chevalier calls the actual expense of making coin brassage, when paid by those who bring the metal to the Mint; he terms any charge made by Government in addition to brassage, seignorage.

It is undesirable in my opinion to make any charge whatever on the manufacture of standard bullion into standard coin, because the actual par of exchange depends, as the "Times" points out, upon the terms upon which standard bullion can be converted into standard coin. I give the extract from the City Article, 21st March, 1870:—

"The change in the method of stating assays of gold bars at the Bank of England is of importance. Hitherto, it appears, they were stated at 1-768th ($\frac{1}{8}$th carat grain) fine, and the importer ran the risk of losing £1 in every £768, while on the average the loss actually amounted to £1 in every £1,536. The new system states the fineness to 1-3 millième, giving a risk of £1 in £3,000, or an actual average loss of £1 in £6,000. The difference between these averages is 0·493 pence ($\frac{1}{2}$d.), equivalent on standard gold to 10s. 6d. for every £1,000. An experienced writer on the subject points out that, small as this difference seems, it nevertheless raises the actual pars of exchange between England and other countries by $\frac{1}{2}$ per mille in our favour. The true par of exchange results from the respective fine gold contents of the coins; the actual par depends upon the terms upon which gold bullion can be converted into such coin. In perfecting these terms, not only does the importer of gold receive more sovereigns for bullion, but on all exchange transactions now current, whether in the shape of bills or contracts of all kinds (foreign investments in State debts, &c., included) an advantage of 1s. per cent. accrues to the country—a considerable item when calculated on the many millions involved. The change is of still greater importance for the future. It is tantamount to relieving the importation of bullion of a tax of $\frac{1}{2}$d. per oz., and to doing away with a premium of $\frac{1}{2}$d. per oz. in favour of exportation. In regard to the French exchange, for instance, whereas now the rate of 25f. 10c. per £1 leads to the export of bullion from the Bank, it must in future fall to 25·08$\frac{3}{4}$ before withdrawals become practicable, and, instead of receiving gold from Paris,

K

when the exchange is in our favour at 25·35, we shall already receive at 25·33¾. The same rule applies to other exchanges; we shall receive gold sooner, and keep it longer."

Now I apply the same reasoning to the 1½d. per ounce which the Bank now levies as an import duty on all standard gold brought to it for sale, amounting to 1·605 per mille. The fact that the Bank receives this import duty instead of the Mint, does not alter the nature of the tax at all; the importer must pay it in order to get sovereigns; it remains a mintage still. (See the Gold Coinage Controversy, page 211).

I think that one of the advantages to be derived from the system of issuing-establishments, which I have referred to, would be the doing away with all charges whatever on the actual manufacture of standard coin, that is to say, that Mint-notes would then be exchanged for standard bullion on delivery at £3 17s. 10½d. an ounce. I here refer the student to Mr. Hankey's "Treatise on Banking," page 91, for a confirmation of Mr. Ernest Seyd's statement, that the Bank of England note-issue is very expensive and wasteful; the assaying of gold for coinage purposes costs exactly double what it ought to do, because the Bank of England does not accept gold bars until they are re-melted into Bank of England shapes, and each bar must be assayed by treble assay at a cost of 4s. 6d. per bar; yet when these bank-bars are sent to the Mint for coinage, the resident assayer must assay them over again, as no metal, which has not been standarded at the Mint, can be melted into the Mint ingot iron forms, which are 1 inch thick by 1⅜ wide, and 24 inches long, and which turn out ingots weighing about 320 ounces.

Mr. Roberts, chemist of the Mint, describes the process of gold assaying thus:—

1st process: the portion of metal to be assayed is adjusted to an exact weight by cutting and filing.

2nd process: the accurately weighed portions of alloy are added to molten admixtures of lead and silver, contained in porous cups, or "cupels," of phosphate of lime, which are arranged in rows in a muffle, or small oven. The proportions of the latter metals are calculated so as to bear a definite relation to the supposed amount of gold and base metals present in the alloy. Result: the lead oxydizes, and is absorbed by the porous "cupel," together with the copper and other oxydizable metals, and the silver and gold remain in the form of a button, which may also contain platinum, iridium, or metals possessing similar properties.

3rd process: the button is reduced by rolling to a thin strip, which is annealed and bent into a loose coil, or "cornet."

4th process: the "cornet" is placed in nitric acid of the specific gravity of 1·25, and the acid is maintained at an incipient ebullition for fifteen minutes; the coil is then treated in a similar manner with nitric acid of specific gravity 1·4; the silver is removed by the action of the acid, and the gold remains in a spongy state.

5th process: the sponge of gold retains the original form of the coil, but it is necessary to impart a certain degree of coherence to the metal by annealing it at a dull, red heat. I may here observe that a small quantity of silver is invariably retained by the gold. It is necessary, therefore, to make check assays on pure gold, or on standards of known composition, upon which the accuracy of the result will in a great measure depend.

6th process: this consists in weighing the gold "cornet." The weights employed bear a decimal relation to the original weight of the assay piece operated upon, and the amount of gold therefore present in the alloy is at once indicated without further calculation.

The method of silver assay by cupellation possesses great antiquity, and is still retained at the Mint. The process consists in melting the alloy in a "cupel," with an amount of lead which bears a definite relation to the weight of the supposed amount of base metal associated with the silver. The weight of silver which resists oxydation at once indicates the amount of precious metal originally present in the assay piece. (First Mint Report, page 103).

209. What is the natural tendency of bullion?

It is to flow towards those markets where the best price can be obtained for it, *so that the raising of the rate of interest to a very high point in any country has the double effect of preventing the exportation of the bullion or coin which is in that country and of attracting bullion or coin to it from foreign countries;* what is called the comparative exchange between any two places in different countries is the price of gold or silver in the one place, as deduced from the price in the other; the data for these calculations are the price in the foreign place, and the rate of exchange at which either the cost is drawn for, or the proceeds of the sales are remitted for, with the relations between the weights and the degrees of fineness at which the prices are reckoned. For instance, we read in the papers, under the head of "comparative exchanges," such statements as the following :

The quotation of gold at Paris is 2 per mille premium, and the short exchange on London is francs 25·25 per £1. On comparing these rates with the English Mint price of 77s. 10½d. per oz. for standard gold, it appears that gold is about one-tenth per cent. dearer in London than in Paris.

By advices from Hamburg, the price of gold is 428 per mark, and the short exchange on London is 13·5½ per £1. Standard gold at the English Mint price is therefore about four-tenths per cent. dearer in Hamburg than in London.

The course of exchange at New York on London for bills at 60 days' sight is 109¼ (25th July, 1872); at this rate there is no profit on the importation of gold from the United States. When the exchange is expressed in paper currency, the premium on gold is added.

PARIS AND LONDON.

? shillings = 1 oz. troy.
1 oz. troy = 31·103496 grammes.
1000 grammes = 3151·30 francs.
1000 francs = 1002 (premium included).
25·25 francs = 20 shillings.

$$\frac{\cdot 311035 \times 3 \cdot 15130 \times 1002 \times 2}{25 \cdot 25} = 77 \cdot 792$$

or £3 17s. 9½d. Gold is therefore 1d. dearer in London than in Paris, or rather more than a tenth per cent., if we take a hundredth part of the given price as one per cent.

HAMBURG AND LONDON.

? shillings = 1 oz. troy.
82 oz. standard gold = 10 Cologne marks, fine (in gold).
1 Cologne mark = 428 marks banco.
13·5¼ marks banco, } = 20 shillings (see Course
or 13·34375 decimal } of Exchange, 203).

$$\frac{10 \times 428 \times 20}{82 \times 13 \cdot 34375} = 78 \cdot 23156$$

or £3 18s. 2¾d. as the price required; so that gold is 4¼d. dearer in Hamburg than in London, or rather more than four-tenths per cent. The Cologne mark, fine, is equal to 3,608 English troy grains.

New York and London.

? United States dollars = £1.
1869 sovereigns = 40 lbs. troy.
1 pound troy = 5760 grs. British Standard.
24 grs. Br. St. = 22 pure gold.
9 gs. pure gd. = 10 gs. Unit. States Std.
258 gs. U. S. St. = 10 dollars.

$$\frac{40 \times 5760 \times 22 \times 10 \times 10}{1869 \times 24 \times 9 \times 258} = 4 \text{ dollars 86 cents.}$$

A pound sterling is taken at 4 dollars 84 cents., according to tariff.

I give the following extract from "Bullion and Foreign Exchanges," page 347, by Mr. Ernest Seyd:—

"According to an old United States' coinage, the dollar was formerly valued at 54 pence, so that 40 dollars were worth £9, and 100 dollars £22·500, but in the present coinage 100 dollars are worth only £20·5483, which makes a difference of 9·4981, or 9½ per cent.; so instead of adopting the new form of exchange, the old valuation of 100 dollars, equal to £22 10s., has been retained; and the difference is quoted by way of premium on the 100. The exchange between New York and London is thus quoted at New York either at a premium varying from 8½ to 10½ per cent., or it is quoted simply—London, at 108½ or 110½. The actual Mint par being, according to this system, exchange on London 109½ dollars in gold, for £22 10s. There is, therefore, no profit on the importation of gold from the United States at the rate given."

210. What generally regulates the rise above or the fall below par, which is constantly going on within the limit which we previously mentioned, between any two places in different countries?

The proportion which exists between the number of bills on either of those places which chance to be on sale in their respective markets, and the demand for their employment. If, for instance, there be many bills on London for sale at New York, and but little demand for them, bills on London will fall at New York till they reach the specie-importing point, that is to say till they fall to what it would cost the owners of them to get them cashed in London, and in ordinary times they will fall no lower; on the other hand, if there be few bills on London for sale at New York, and great demand for them, bills on London will rise at New York till they reach the specie-exporting point, that is to say till they rise to what it would cost to transmit bullion or coin instead of making use of bills, and in the general run of instances they will not rise above this point. The price of bills, therefore, depends upon supply and demand, like the price of any other merchandise, but it does not generally rise higher or fall lower than a certain point on either side of the par of exchange, as above or below this point the demand for bills ceases, and specie is imported or exported, as the case may be.

211. What other element affects the price of bills?

The credit of the drawer, acceptor, and indorsers of the bills. A bill backed by first-class names will always command a higher price than a bill backed by second-class names; anything which tends to diminish the credit of any country will at once lower the price of all bills drawn upon it.

212. In what case may the price of bills fall below the specie-importing point?

In any case of sudden panic where the holders of the bills cannot afford to wait sufficiently long to get the bills cashed in the country they are drawn upon. The capitalists who buy the bills under such circumstances make so much extra profit, but they must lose the use of the money which they give for the bills during the interim.

213. In what case may the price of bills rise above the specie-exporting point?

In any country where the currency is much depreciated from over-issues of paper money, and where the export of bullion is forbidden by law, the price of bills may rise much above the specie-exporting point whenever the merchants of that country have no produce of any kind available for export as a means of paying their debts.

214. What are bills drawn in blank?

Bills which are not based upon any actual trade transactions between any two different countries, but which are either drawn in anticipation of the ordinary trade transactions between those countries or are drawn as a means of obtaining the money which is paid as their price; in either case they may be called foreign accommodation bills. They have an important influence on the state of the foreign exchanges, as there is generally a large amount of them afloat.

215. When the exchange between London and New York is favourable to New York, and bills on London are near the specie-importing point, what effect has this on the exporting and importing merchants in New York?

The importing merchants have a decided advantage over the exporting merchants, for they can buy the exporter's bills at a discount, and thus pay for their own importations.

216. What effect has the same state of the exchange on the exporting and importing merchants in London?

The exporting merchants have a decided advantage over the importing merchants, as they can sell the bills which they draw on New York at a premium, while the importers must either pay a premium for the bills they want or be at the cost of transmitting coin or bullion.

217. What is the only means which any particular country possesses of regulating the foreign exchanges, as far as it is itself concerned, supposing that its currency is not depreciated in value from over-issues of paper money, or a base coinage?

The raising or lowering the rate of interest,

according to its state of debt or credit with all other countries—

(1) When it owes more to other countries, taking all its transactions together, than they owe to it, an efflux of coin or bullion soon takes place, and the rate of interest must then be raised at home ; this will have the effect of restricting its purchases in foreign countries and increasing its sales at home for exportation, and the foreign exchanges will very slowly become more favorable.

(2) When other countries owe more to it, taking all their mutual transactions into account, than it owes to them, an influx of coin or bullion soon takes place, and the rate of interest will then fall of itself at home, this will have the effect of increasing its purchases in foreign countries, and restricting its sales at home for exportation, and the foreign exchanges will slowly become less favorable. The state of the foreign exchanges, as far as any such country is concerned, depends upon its state of debt or credit with all other countries.

CHAPTER VII.

DEPRECIATION OF THE CURRENCY.

218. What is meant by the term depreciation of the currency?

A diminution in its purchasing power. Whatever substance may be used as currency, an excessive quantity of it (more than is required by the wants of the community) necessarily causes a diminution in its purchasing power.

219. Could a currency then, consisting of the best gold coin only, be depreciated?

Certainly; provided that the exportation of gold could be *altogether* prevented, the amount of currency in use would soon become greater than what was required by the wants of the community, and its purchasing power would diminish in the same proportion.

220. What prevents a properly regulated currency from being depreciated?

A very large proportion of such a currency possesses an intrinsic value, and can be exported to other countries whenever it is in excess of the exact amount required by the wants of the community to whom it belongs.

221. What is the principal cause of depreciation in currencies at present?

The age for uttering base coin by Government

authority is gone by, and a more polite way of obtaining money is at present in fashion, viz., unlimited issues of Government bank-notes which are inconvertible, and at the same time a legal tender. We use the term good coin in this chapter in contradistinction to base coin, and as applicable to either gold or silver money whichever be the standard of value chosen.

222. On what does the purchasing power of such notes depend?

It depends on the quantity of them that is issued and remains in circulation, and however wealthy the Government may be that issues them, any large increase in their amount soon produces a diminution in their purchasing power.

223. How can the extent of this diminution be shown?

By the difference in the price of any article when paid for in good coin, or in inconvertible paper notes, or in other words by the relative value of the inconvertible paper when compared with the standard money of the country, whether it be gold or silver coin. Thus, in America, at present 114 inconvertible legal tender paper dollars are required to purchase what 100 gold dollars can readily purchase, while in Austria from 112 to 130 legal tender inconvertible paper gulden may be required to purchase what 100 silver gulden can purchase in the same market.

224. In what way then must we look on an inconvertible legal tender paper currency?

As a distinct and independent substance

made use of as currency; gold and silver under such circumstances assume more and more the character of merchandise, while they lose the character of money.

225. Does a depreciation of the currency, arising from over-issues of paper money, affect all prices uniformly?

It does, but we often cannot see its effect, for each commodity is affected by its own particular circumstances, and its price in paper money will rise or fall accordingly.

226. Is it ever necessary for States to resort to a suspension of cash payments, and to inconvertible bank-note currencies?

I agree with Mr. Ernest Seyd when he states, "That over-issues of bank-notes are thoroughly legitimate, when the pressing necessities of a nation by war or rebellion cannot be met otherwise; then they must take place. Indeed, without such paper-money, the United States would have been unable to repress the late rebellion, and their issue became an imperial duty." (Metallic Currency of the United States, page 100). The celebrated "green-backs" were the unfunded debt of the American people; and when the citizens bought Government stock with their green-backs, which were then burnt, they added to the funded Debt that portion of the unfunded Debt, which they withdrew from the circulation.

227. What is the principal inconvenience caused by over-issues of inconvertible paper-money?

The fluctuations which are constantly taking

place in its value, and the greater the fluctuations the greater is the inconvenience which is felt by business men, for the debts which they contract in a currency of one value they have to pay in a currency of either a higher or a lower value as the case may be. Either the buyer or the seller is sure to lose as the paper currency becomes more or less valuable.

228. On what does the amount of currency in use in any country depend when that currency is not depreciated, and when the foreign exchanges are either at par, or but slightly affecting the circulation?

It depends on the exact amount which the wants of trade require in that country; its Government may determine what proportion of this amount shall exist in the form of bank-notes issued against other security than that of good coin, or bullion, and what proportion shall exist in either bank-notes issued against good coin, or bullion, or in good coin itself, but further than this they cannot go, for if they once attempt to increase the amount of currency above what we may call its natural limits, by extra issues of paper-money, a diminution in its purchasing power will infallibly take place. They could not increase the total amount of currency in any other way but by over-issues of paper-money, if we except the uttering of base coin, for it has been found impossible to prevent the exportation of gold or silver, and any good coin, which was not required at home, would soon be exported. Let us suppose ninety-nine millions sterling to be the exact amount of currency which the wants of trade require in some particular country, when the foreign exchanges are but slightly affecting the circulation, then the Government of that country might, if they chose, have sixty-six millions of this amount in bank-notes issued against other security than that of bullion, or good coin, and as long as the same state of the foreign exchanges continued, this currency need not necessarily be depreciated, and the bank-notes might be apparently convertible.

229. What would be the danger then in following such a system?

If the foreign exchanges became unfavourable, and gold or silver were required for exportation, there would be two-thirds of the circulation of that country in the form of promises to pay certain quantities of the precious metals, and but one-third of the actual substance required for payment to meet all the promises.

The bank-note circulation of the Bank of France is about ninety-seven millions of pounds sterling (20th July, 1872), with a metallic reserve of about twenty-nine millions, and with permission to issue more bank-notes to the total amount of one hundred and twenty-eight millions (Law of 15th July, 1872.) The very large metallic circulation in use before the war of 1870, has rendered it possible for M. Thiers to increase the bank-note circulation of the Bank of France by about thirty-eight millions of pounds sterling, without subjecting them to a greater depreciation than about one-and-a-half per mille, July, 1872. The object of this great statesman should now be to give the Bank every possible support in the policy which it has hitherto been following, of diminishing its loans, and thus of restraining its issues as much as possible. M. Thiers should also give every possible facility to production in France, so as to enable his people to supply themselves again with the precious metals.

230. What is one of the errors into which the opponents of the present currency principle often fall?

They forget that we have chosen gold as our only standard of value in Great Britain, and that prices are consequently but the quantities of gold which each article can command when exchanged. Mr. Macleod, in his excellent work on the 'Theory and Practice of Banking,' speaks of the great banking fallacy of the present day, that the issues of banks should be rigidly restricted in amount to what would be the amount of gold-coin if they were not in existence; a little further on, he says, "if, therefore, when to a previously existing gold currency there is added a paper currency, if new fields of industry are opened up, if new manufactures are started, if lands are reclaimed, they will absorb a currency far greater in amount than would have been sufficient if they had not been attempted, and so operations may be generated of a far greater amount than if there was nothing but a gold and silver currency. Now it seems so obvious a truism, that, if while any given amount of currency be required to conduct the operations of a country, and any given increment be made to that currency, then if the increased operations which that increment of currency gives rise to, are in the same proportion to it, as the operations previously carried on were to the previously existing currency, the value of the whole currency will not be altered, that it seems incredible that it should be required to enunciate it, and moreover that it should be *fatal to the leading banking principle of the day.*" The value of the whole currency under such circumstances might not perhaps be altered if estimated in any commodity but standard gold, but the paper part of the currency when brought to this touchstone could not maintain a precisely equal value with that exact amount of standard gold which it purported to represent for this simple reason, that when gold became wanting for exportation there would be more paper notes in circulation than could by any possibility be converted into their equivalent in gold, and promises to pay being then easier obtained than actual payments in gold, the paper currency would be depreciated, however much it was needed by the operations of trade. If the gold discoveries in California and Australia had never taken place, the trading operations of the world would certainly have gone on increasing, though at a far slower rate; gold would have become slowly more valuable and prices would have fallen all over the world; nor could gold under such circumstances have been prevented slowly rising in value by any issues whatever of *really* convertible paper money! We might just as well say that gold could be prevented from falling slowly in value, when the civilised world have as much gold and silver as they require for their trading operations. Mr. Macleod has defined convertibility admirably, "so long as the market or paper price of gold bullion coincides with the mint price, it is an infallible proof that the currency is not depreciated—that paper is at its par value." How long would this be the case if Lord Overstone's great law was not in force, viz., "*that the sole duty to be performed in regulating a currency composed of bank-notes and coin is to make its amount vary as the amount of a currency exclusively metallic would vary under the same circumstances, because the mere obligation to pay in gold on demand is not a sufficient security to the public that the power to do so will at all times be maintained by the issuers of bank-notes.*" We have previously remarked that the wants of trade prescribe the amount of currency which any country requires, as long as that currency continues undepreciated; now it is the proportion which exists between that amount of currency which is represented by bank-notes issued against other security than that of gold coin or bullion, and that amount of currency which is represented by gold-coin itself, or bank-notes issued against bullion, supposing gold to be the standard of value, which must be looked to by the *true* advocates of convertibility as defined by Mr. Macleod; if this proportion is not carefully attended to, the total amount of currency may not be at all greater than the wants of trade require, and yet the paper money will not preserve its convertibility, if put to the test by gold being required for exportation; *it simply cannot if its amount is much higher than the amount of gold coin or bullion which chances to be in the country.* The 'Economist' says with great truth, that the authorities *even* of the famous system of monetary philosophy known as the currency principle no longer insist upon the variations of the bank-note circulation as the symptoms to be chiefly regarded; but *why* is this the case? Simply because the act of 1844 has succeeded in maintaining Lord Overstone's great law, and *therefore* rendered the variations which can take place of no importance; a diminution in the authorised bank-note circulation of some of the banks in the United Kingdom may not be for the interest of those banks, but can *hardly* be a source of much inconvenience to the public as long as the Bank of England has the power of issuing against bullion, and all the old banks in Scotland and Ireland have the power of issuing against sovereigns. "Plans for an improved system of currency," says Lord Overstone, "are frequently laid before the public; the exclusive object of these systems is to obtain for the paper currency to be issued under them a greater degree of security, than that which is supposed to attach at present to the notes of the Bank of England. This end the authors of these schemes generally propose to accomplish by contrivances, which they deem to be extremely ingenious, but which always resolve themselves into *the simple plan of making property of some kind or other the basis of the circulation.* Sometimes the plan suggested proposes to issue a paper currency against the security of land, sometimes against the security of the public debt, and sometimes against merchandise in the docks; but having provided for the security of the notes, the plan generally terminates at this point; the projector apparently conceiving that he has satisfied all the desiderata of a good paper currency, although he has introduced *no specific measure for regulating the amount of that currency, and maintaining its value relatively to the currencies of the other countries of the world."*

231. What does the "Economist" declare to be the true and complete doctrine in this matter?

It declares that the *entire* credit currency ought to enlarge and contract with the influx and efflux of bullion, and that the rise or the reduction of the rate of discount is the effectual and sole mode of effecting either that enlargement or that contraction. In the present day, says the same journal, it may be safely laid down that if you can keep the rate of interest right, so far as the exchanges go, and so far as prices are concerned, *you may leave bank-notes alone.* If you secure that the wholesale currency in which all great loans are effected, and in which all large bargains are settled, shall be kept right, you need not care for the retail currency in which petty loans are made, in which clerks' wages are paid, and small transactions brought to a termination.

232. How does this differ from the doctrine that Lord Overstone lays down?

"I consider," says that scientific banker, "the *money* of the country to be the foundation, and the bills of exchange, &c., to be the superstructure," having previously defined *money* as the quantity of metallic coin, and of bank-notes promising to pay the coin on demand which are in circulation in the country. This is attaching a very different importance to what the "Economist" designates as "the retail currency" of this country. Lord Overstone's principle, with regard to the management of the currency is this, *that the amount of money in use in the country, taking his own definition of the term, should be made by law to vary when composed as it now is, exactly in the same way as it would vary if composed of coin only.* There is *no means* of maintaining the absolute and constant convertibility of our bank-notes but by strictly limiting the amount of them that may be issued against other securities than those of gold coin or bullion; should we require bank-notes above this amount on account of their superior economy and convenience, such notes should be but so many certificates of the deposit of a corresponding amount of gold coin or bullion. "The Act of 1844," says Mr. Weguelin, "does not restrain the circulation in the hands of the public; it simply says that if the public employ more circulation than they do at present, it shall be on the security of gold." The difference between bank-notes under our present currency system and auxiliary currency is this, that the amount of bank-notes which can be now issued over and above a certain fixed amount, is made by law to depend on the amount of gold coin or bullion which the issuing banks may chance to get hold of; there is, therefore, an artificial regulation affecting bank-notes, which does not affect cheques, bills of exchange, &c. Now, if we remove this regulation, the 'Economist' would be right in classing coin, bank-notes, and all auxiliary currency together; nor would there then be any other means of regulating "the entire credit currency" of the country but that which it indicates.

233. Where, then, is the error?

It is in the fact that raising the rate of interest is the very way to get "the retail currency," which, by the "Economist's" hypothesis is to be left alone, into such a disordered state, that " the wholesale currency" would soon be affected, and the entire credit currency of the country would collapse. Lord Overstone says, "I will attend to the money which I consider to be the foundation, and I will allow the superstructure to take care of itself." The "Economist" says, " We will attend to the foundation and the superstructure both together, for we do not recognise the distinction between bank-notes and auxiliary currency, which the theorists of 1844 believed that they established." The gold discoveries in California and Australia have nothing to say to the matter, for a bank-note says nothing about *the value* of the pounds which it promises to pay; the true question is how can we best render it a certainty, that the public can *always* change the bank-notes which they hold into the actual sovereigns when they want them, without any loss or risk whatever. According to our present system, if gold should be depreciated from over supplies, our circulation (however small our excellent credit system may have rendered it) will expand itself in exact proportion to that depreciation.

234. What does the 'Times' say?

It would be hard for *any one* to explain by what principle the Government is called upon to interfere with the freedom of individuals to put forth whatever pecuniary promises they may see fit, whether in the shape of notes or bills, or anything else. All that is necessary is, that no notes should be a legal tender except those issued by the Bank of England, &c. Notes of all other descriptions the public might be left to issue or to receive at pleasure, the issues being, in cases of fraud, amenable to the ordinary law of the land (City Article, 13th February, 1865).

235. What would be the effect if the views here expressed were carried out?

We should probably obtain one national state bank for issuing purposes more quickly than with our present system, for men of business never could stand such a mixture of paper currencies as would then be in circulation. People cannot appeal to the laws of the land in cases of fraud of this description; men of business have not the time to inquire into the solvency of each bank whose notes are in circulation; it ought to be the duty of Government to take care that the business of issuing bank-notes is in proper hands, and where they did not do so, as in the United States, wild-cat paper and forgeries were so common, that men of business had to keep dictionaries of bank-notes by them, which came out monthly, containing the note-marks, and all the known forgeries, and before they took the notes they had to look them out.

The American Government has now adopted a system of national banks, which are allowed to issue bank-notes received from Government, provided that they give as a guarantee Government securities, worth one-tenth more than the bank-notes which they receive, and that they put one-third of their subscribed capital in Government securities. They must also keep from 15 to 25 per cent. of their liabilities (bank-notes and deposits combined) in greenbacks. No other banks in the United States are allowed to issue bank-notes.

236. Can you mention *any one* whose opinion is different from the " Times" ?

" The claims of right to such freedom of action in banking ought to be strenuously resisted; they do not rest, in any manner, on grounds analogous to the claims of freedom of competition in production."—Mr. Tooke.

" We do not want," says Sir Robert Peel, " a considerable quantity of bank-notes at the lowest price possible, but we want a certain quantity of bank-notes whose value is exactly equal to the gold which they represent."

237. What may we conclude was Sir Robert Peel's ultimate object with regard to the currency system?

To get rid of all the anomalies and difficulties which must always arise from the business of banking being mixed up with the business of issue, by creating one national state issuing establishment, and depriving all banks of the power of issue. " It has always appeared to me," says Mr. Hubbard, M.P., " that if it is the proper function of the State to provide the people with currency, it ought not to be limited to one material more than another; if it is the duty of the State to provide a *safe* currency of gold, silver, and copper, why not also of paper? If this question is ever to be brought to a satisfactory issue, Parliament must affirm the right of the State to create all money, and be the coiner of all expressions of value."

CHAPTER VIII.

LAND.

238. Is land to be looked upon as capital?

Once land is occupied and appropriated it becomes capital, because it is then employed with a view to increasing our capabilities of production by the material profit derived from that employment. If a man has five thousand pounds in ready money to invest as capital, he may lay it out in land, or in a manufactory, or in a ship, or in a mine, but whatever be the way of laying out his capital which he may select, what he obtains is equally capital whether it be a piece of land or a brewery.

239. What gives land its value?

The real or fancied advantage derived from the use to which it is put. Land may derive its value from agriculture, from buildings, from mines, from situation, etc., and the value of land will be dependent upon the proportion which exists between the supply of it to be obtained in any particular market, and the demand for it in that market. In new countries, where there is much unoccupied land, like the United States or Australia, the supply is large in proportion to the demand, and land may be bought almost for the mere cost of fencing and appropriating it; in old settled countries like

England and Scotland the demand is very large in proportion to the supply, because many wealthy people wish to acquire land as a means of increasing the consideration in which they are socially held, and they are therefore satisfied with but a small return for the investment of their capital; in the one case land is sold below its value as a productive engine, in the other it is sold very much above it.

240. Is it true that land is a different kind of property from everything else, " that it is a monopoly existing in limited quantity and not susceptible of increase?" (Mr. J. S. Mill).

The acquisition of land is perfectly free, nor can its possession be in any way termed a monopoly, for there is always plenty of land in the market for those who have money to buy it. Cultivated land might, perhaps, be said to exist in limited quantity, but then it is largely susceptible of increase; if land was really a monopoly, all other commodities would be monopolies on the same principles. Take any one commodity, hats for example. The limit to the production of hats is the existing number of human heads requiring hats; if A be the number of heads, we can but want A hats; if we produce A^2 hats, we cannot get rid of them without a fall in price.

Let us now compare hats with land, and we shall see that the increase of occupied land is what is principally required to increase A or the number of heads requiring hats because we can only permanently increase our numbers by increasing our means of subsistence. If our world was all occupied and cultivated like a garden—if land was indeed a monopoly existing in limited quantity, and not susceptible of increase, the production of hats would be also a monopoly—hats would then exist in limited quantity not susceptible of increase. As all commodities are produced to gratify the wants of mankind, the limit to the production of any particular commodity may be reasonably laid down when a supply sufficient to gratify the wants of all who require it at its lowest producible price, is brought into the markets of the world.

Whatever be the commodity, its production with a sufficient profit to the producers, must be limited by the number of those who want it. Had the Greeks been able to persuade other nations to eat currants in the same way that we in England eat them, the production of currants would have been a source of great wealth to Greece, but they forgot that the production of currants must be limited by the number of those who want currants; and when they extended their currant vineyards some years ago, without increasing the number of their customers, they only brought down the price of currants.

The human race cannot increase permanently in number without increasing their means of subsistence, and there are but two ways of doing this, either by getting more produce from the land already under cultivation, or by reclaiming new and untilled land. This last work has been facilitated greatly by improved means of communication and increased facilities of transport. The last custom-house returns show how much the English people depend upon foreign supplies of grain; in the first half of the year 1872 there were imported into the United Kingdom 15,636,842 cwt. of wheat, more than half of it coming from Russia; 1,446,084 cwt. of wheat-meal and flour; 7,036,697 cwt. of barley; 5,632,269 cwt. of oats; 454,855 cwt. of peas; 1,622,952 cwt. of beans; 8,181,066 cwt. of Indian corn; and 2,581 cwt. of Indian corn-meal. The declared value of this importation was £19,376,938.

A limited quantity of land, however well it may be cultivated, can produce a limited quantity of produce only, and if land in our time was really limited in quantity, the production of commodities would soon reach its limits too. " Cette limitation de la terre," says M. Baudrillart, "paraît un argument fort contestable en présence de la masse énorme de terres non encore exploitées.

241. Why is new land not taken into cultivation more rapidly than it is?

Reclaiming land is an expensive operation, requiring the outlay of capital, and unless there is a good market for the produce, it does not pay to carry it on. Men with no capital but their labour might starve at such work. Hence land will be but slowly occupied and reclaimed as capital can be applied to it, and capital will not be applied without a reasonable prospect of profit; it will not do to ship cargoes of labourers from old settled countries, and to say to them on arrival in a new country, "here is plenty of wild uncultivated land;" for what are they to live on while they are engaged in reclaiming it? We might nearly as well stick a lot of new spades in the ground and tell them to begin to dig.

Once the land has been successfully reclaimed and occupied, once it is in good order for productive purposes, the wants of the settlers become a source of wealth to their new country, because they have then got something which they can offer in exchange; an increase of prosperity is the result, which soon leads to an increase of population.

242. What should be our object in the management of land as a productive instrument?

We should follow whatever system we find to give the greatest possible amount of produce without injury to the land.

243. What can be said in favour of large farms?

"The power of capital and labour," says M. Rossi, "cannot be properly developed unless these two instruments are applied on a grand scale to great undertakings. A great manufactory will give a net product which is larger than what we could obtain from the same productive powers if divided amongst ten small manufactories. On the one hand the expenses of starting an establishment, and of superintending and directing it, increase in proportion to the number of separate establishments; on the other hand a proper division of labour, and the employment of powerful and expensive machinery, is possible in those establishments only which give sufficient employment to all the different classes of hands, and which are able to produce things on a grand scale. The larger the quantity of goods produced is, the smaller is the share which each separate article has to contribute towards the original outlay in buildings and machinery; if two machines of the same power produce in the same space of time, the one a hundred thousand yards, the other two hundred thousand yards of the same kind of stuff, you may say that the first machine is twice as expensive as the second one, that in one of these undertakings twice the necessary capital is being employed; you may say, too, that as one yard of the stuff in the first manufactory costs the manufacturer as much as two yards cost in the second, the net product of this last is twice as great as that of the first.

"Now can agriculture escape from the doctrines here laid down? We find certain things to be true with regard to manufactories of furniture and goods; is what is true when applied to them, not to be true when applied to manufactories of wheat, hemp, or clover? It is clearly applicable to all manufactories alike. Divide in your imagination one large farm, one great agricultural manufactory, into thirty little undertakings, quite independent one of the other, each with its own farm buildings, its own tools, its own machinery, its enclosed yards, its farm roads, and the special superintendence which it requires, and you will see the cost of production, and especially the capital required for the original outlay, increasing in a most striking way. And remark, besides, that in dividing one farm into thirty lots, we do not suppose any one of the lots to be so small as to exclude the use of machines, or to prevent a plough being worked. If this was not so, if we divided our farm into such small portions that machinery would not be available, the results of the division would have a far greater effect on the wealth of the nation. These little farms, too, retard the progress of agriculture, because they do not attract the capital of skilful and educated men, nor offer inducements enough for their activity; the good methods of farming, which have been already discovered, are rather pushed aside than attempted by these small farmers, with hardly any capital, whose ignorance makes them distrust the advice of richer friends." (" Cours d'Economie Politique ").

244 What can be said in favour of small farms?

We refer the student to the writings of Mr. John Stuart Mill for the answer to this question, small farmers are so much more comfortably off in his books than they are elsewhere; but we shall give an extract from Mr. Mure's account of his visit to Flanders in 1869, as it serves to elucidate the matter. The Cadastre Survey, says Mr. Mure, published in 1856, states that in West Flanders the arable land, amounting to 677,005 acres, is divided amongst 86,225 occupiers, showing the proportion of $7\frac{1}{2}$ acres to each. In East Flanders the division of 545,245 acres among 88,305 occupiers, reduces the mean size of holdings to $5\frac{1}{4}$

acres. The peasant tenant of Flanders is the descendant of a race of peasant-proprietors, who, by their skill and industry, have snatched hundreds of thousands of acres from the dominion of the sea, forged wealth out of pure silica, and converted a howling wilderness into a fruitful garden. The fruitful garden remains, but it has now, under the pressure of competition, to fulfil the threefold mission of land which has arrived at commercial value, viz., to provide rent for the landlord, profit for the tenant, and wages for the labourer. The peasant-proprietor, cultivating his own land, has disappeared, and with him likewise security of tenure. The tendency of the law of testamentary division is not only to break up extensive estates and root out the land monopolist, but also to keep continually forcing a large area into the market in small parcels, within the means of that very numerous class, whose love of land amounts to a passion. As generations succeed each other, and population increases, and with it competition, the proprietor, finding it impossible without excessive labour to extract from his land a fair return for his capital, and unable to resist the temptation of high offers, soon learns to prefer the ease and petty powers of the landlord to the toil of the husbandman. The tenant enters rack-rented, and, with the dreaded rent-day ever hanging over him, is obliged to sacrifice every other consideration to the cultivation of his land. His children, from their earliest years, having to take their share in the daily routine of labour and anxiety, grow up in complete ignorance, and the land, under the high pressure of intense industry, is clothed, fed, and adorned by a lavish use, or rather an abuse of the energies of the over-wrought slaves who water it with their sweat.

Speaking of Flanders, M. de Laveleye, a very high authority, says: "Mais là aussi nous avons été frappés du triste contraste que présentaient ces magnifiques récoltes, et l'existence misérable de ceux qui les font naître. Malheureusement la condition des hommes laborieux qui ont amené l'agriculture à un si haut dégré de perfection n'est point en rapport avec la masse des produits qu'ils récoltent. L'ouvrier agricole des Flandres est peut-être celui de tous les ouvriers de l'Europe le plus mal nourri; le petit fermier ne vit guère mieux, et si l'on y regardait de près, on se convaincrait que, loin de tirer du capital engagé dans son exploitation les 10 pour cent jugés nécessaires en Angleterre, il n'en obtient pas 3 pour cent en sus du salaire qu'il mérite par son travail personnel."

My own experience on the spot has confirmed M. de Laveleye's opinion. I went to the garden of Belgium impressed with Mr. Mill's views, and prepared to find a paradise of agriculture resulting from a peasant proprietary.

I found for the most part naturally miserable soil and marvellous fertility, a curious skill in husbandry and a most lamentable ignorance, complete stagnation without much absolute misery, and a swarm of small proprietors possessing all the vices which too often characterise needy power lording over and quarrelling with a rack-rented and oppressed tenantry.—(Letter to the "Times," December 15, 1869).

245 Is there any country where small farms are advantageous ?

Mountainous countries, where arable land lies in small patches, can only be managed on this system. A large part of Switzerland, for instance, is divided by its natural conformation into small holdings.

246 Is it true " that private persons have been allowed to appropriate the source from which mankind derive their subsistence, that is to say, the land ?"—(Mr. J. S. Mill.)

It is a happy thing for mankind that it is true; land in civilised countries must belong to someone, and it is far better for all parties concerned, that it should belong to a number of individuals who will each give attention to his own property, than that it should belong to the State, which has in most cases quite enough on its hands without such a burden.

Landed proprietors are often represented, even by well educated writers, as a class of men who repose luxuriously in their country seats, while heaven is always showering its choicest blessings upon them, and while their land, without any help from themselves, is ever increasing in value "with the growth of society,"—one would imagine that such men would never consent to sell their land—yet strange to say, one week's advertisements in the "Times," during May, 1872, displayed announcements of the sale by auction of about 62,000 acres in 24 English counties and in Wales, exclusive of many small properties and of building land.

The International Society, says Mr. Fawcett, in his powerful speech in the House of Commons on this subject, proposed that the State should buy up the land, and all other instruments of production at their present price; and then it was said the people would be able to obtain land and houses and means of industrial production at a cheap rate. Now a high financial authority had calculated that if the land in this country were bought, it would require £4,500,000,000. To raise such a loan it might be fairly assumed that the rate of interest would rise; if it rose only to $4\frac{1}{2}$ per cent., the interest on this sum would represent £200,000,000 a year. Even supposing that the land were let at its present price, it would then only realise £150,000,000 a year. Therefore to begin with, as the result of this transaction, there would be a loss to the nation of £50,000,000 a year. But this was only one of the difficulties of the question. There would be a loss of £50,000,000 a year if exactly the same price was charged for the land. But if the same rent was charged for it, the public would be in exactly the same position as before, except that they would have lost £50,000,000 a year by the transaction. But the International Society said, "We hope to get houses and land at a fair rental." The more rents were reduced, however, the greater the deficiency to be made up and the national burden thrown upon the resources of the country. Admitting even that the deficiency was made up from the clouds, what would happen? If the State had the land, and let it at less than its present value, how could the bargain be adjusted? Who would be the favoured persons to have land in favoured situations? If the land were let at a uniform price, who would have the rich fertile lands close to the large towns, and who would be relegated to the barren moors of Yorkshire, or the heaths of Devonshire? If the land was not let at a uniform price, it was obvious that you would only bring into operation the same competition as now existed; you would place in the hands of Government an enormous and unprecedented amount of patronage, an unequal power of rewarding supporters and punishing opponents; and under such a blighting influence England would not exist as a nation even for one generation.

247. Is there any truth at all in the assertion that land exists in limited quantity, not susceptible of increase?

There is when you specify the particular land to which you refer; land, for instance, in England certainly, does exist in limited quantity.

248. What is the result of this fact?

Simply that the supply is small in proportion to the demand. Wealthy people, who want to obtain a good social position, choose to pay more for land in England than it is worth as a productive engine; capital invested in land there makes a less return to the purchaser than if invested in other ways. The farmer who rents this land is in a far better position than the proprietor who lets it. It is nothing to the farmer what the proprietor gave for the land; he calculates what its producing powers actually are, and offers a rent for it in proportion to those powers. Were the farmer to be made a proprietor of his farm, either he would be ruined by the increase of rent required to enable him to complete the purchase, or the State would have to advance money to him at the general loss of the nation. A farmer can make say 10 per cent. by judicious farming in England, an investor of capital in the purchase of land can make perhaps 3 per cent. by his purchase.

The problem which the members of the Land Tenure Reform Association have to solve, is how from such premises as these they can profitably create a small proprietary in England and Scotland. Mr. Gladstone put the case well before us when he said, that the tenant in England knew perfectly well that he had a much more lucrative return for his capital by using it as a tenant-farmer than as a proprietor; and in England instead of tenants becoming landlords, small proprietors sold their land, and farmed on a large scale as tenants ("Times," 16 May, 1870).

249. What effect has the gradual adoption of the large farm system upon the agricultural population?

The census returns show that agriculture is supplying less and less employment, as machinery is introduced, and that the population in the rural districts is therefore gradually diminishing in numbers, while the population in the towns is increasing.

250. What is rent ?

Rent is the price paid for the use of houses, land, etc. The rent of land varies for the same reason that the rent of a house varies ; houses pay a rent proportioned to the advantages which their possession offers, whether they be advantages of position or accommodation ; land pays rent according to the return which it is able to make to those who manage or cultivate it, of whatever nature that return may be.

251. What was the origin of rent ?

An increased demand for land, as a productive engine, led to rent being paid for its use :—
" Die Boden-Rente kommt von der durch das Anwachsen des National-Wohlstandes vergrösserten Nachfrage nach Grund - Stücken." — (Max Wirth, " Grundzüge der National-Oekonomie ").

252. Explain this by an instance.

Some years ago English gentlemen could go to Norway, to fish for salmon in the rivers there, without having to pay for their sport. The rivers in Norway are now let to sportsmen, just as they are in Scotland. The rent which is paid for this fishing is demanded for the very same reason that rent is demanded for land. The proprietors of the rivers discovered that they possessed a productive instrument in the right to the fishing, and they demanded a rent, which the fishermen were glad to pay, in order to secure the fishing for themselves. A really good river, famous for fishing, will let at a much higher rate than a river with but few salmon in it, and land is let on exactly the same principles; a farm which is very fertile and productive will command a much higher rent than a farm consisting of poor, barren soil. Rent does not arise from one soil being better or worse than another; it is the price paid for the use of a productive engine, and the price will be high or low according to what the engine is able to produce.

253. Does rent, then, form a part of price, or not?

Half the ponderous disquisitions on this subject might have been avoided, had economists remembered that price is a vague term, and may mean either the particular price of a given commodity in one market on a given day, or the average price of that commodity in any country, during a given number of years. If we take the first meaning of price, rent need not necessarily form a part of price at all. No one, when they buy the produce of land, ever inquires what rent the producer paid for the land; that is nothing to them; they buy as cheap as they can, and circumstances may make the produce so cheap that there is nothing left in the price out of which rent could come. The producer, in this case, must pay his rent from other sources, and rent does not form a part of price. We may here observe that this is alike true in certain cases of all the ingredients which combine to make up the cost of producing any commodity. When the fashionable world suddenly gave up wearing brass buttons, the price of brass buttons had nothing to say to the cost of producing them; even the metal itself, in such a form, had a very trifling value. Whatever be the instance given, we may find the same result on certain occasions. A coat of woollen cloth may be bought without the slightest reference to the price of wool; a pair of shoes, without the shoemaker's wages being necessarily considered, or the price of leather.

The result is very different when we speak of average price during a given number of years; here the cost of production comes in simply because the producer would not continue to produce the commodity unless it paid him to do so. In the case of the brass buttons, their manufacture was at once discontinued; and in any case we can bring forward, the average price of a commodity will be regulated by the cost of its production, including in this term the average rate of profit required by the producers, in order to pay them for their labour. Hence, if we take the average price of beef, mutton, or butter, rent forms a part of their price just as much as a herdsman's or dairymaid's wages do. Mr. Mill asks the question—"Does rent enter into cost of production?" And he says, "the answer of the best economists is in the negative." But in the same page, he adds, "no one can deny that rent sometimes enters into the cost of production. If I buy, or rent, a piece of ground, and build a cloth manufactory upon it, the ground-rent forms legitimately a part of my expenses of production, which must be repaid by the product. And since all factories are built on ground, and most of them in places where ground is peculiarly valuable, the rent paid for it must, on the average, be compensated in the values of things made in the factories" ("Principles of Political Economy," People's Edition, p. 284).

Alter these words to suit our view, and we have what is an equally true statement:—"If I buy or rent a piece of ground and turn it into a dairy-farm, the ground-rent forms legitimately a part of my expenses of production, which must be repaid by the butter I produce. And since all dairy-farms consist of ground, and most of them are to be met with in places where ground is peculiarly valuable, the rent paid for a dairy-farm must, on the average, be compensated in the value of all butter, produce, etc., etc., made on such farms."

254. Is there any other element always affecting rent?

In countries like Russia and the United States, where there is much land still uncultivated, there is the continual operation of bringing new land into cultivation; this, as it is always tending to increase the supply of certain kinds of produce, also tends to diminish their value, and so to keep rents low. If the rent of land was not much lower in the Western States of America than it is in England, the Western farmers could not export the thousands of barrels of flour which they send to us, because wages are higher with them than they are with us. The quantity of wheat which we now import from Russia is a proof that each year more and more of the vast steppes above the Black Sea are being brought into cultivation. In this age of free exchange those who, for a continuance, can give us large supplies of any commodity cheaper than anyone else can, are the regulators of its price; anyone who attempts to produce the same commodity under more unfavourable conditions may lament his hard fate if he likes, but sell he must, if he means to sell at all; for no one cares what it cost him to produce the commodity, and they know what they can get it for elsewhere.

255. If there was no such thing as rent, would the produce of land be cheaper than it is?

We cannot suppose rent to be done away with altogether, without also supposing an unlimited supply of fertile land being within easy reach of everyone who might wish to appropriate it; and in such a condition of society, which existed at a pre-historic period, the produce of land was, no doubt, much cheaper than it has since been. Taking the question as an economic crux, we may observe that the cheapening of agricultural produce can always be traced to those countries where the rent of fertile land is the lowest. As rent, in the long run, forms a part of the produce of land, the getting rid of rent would at first increase the producer's profits, but competition amongst producers would slowly bring profits back towards their former level, and in the end give consumers the benefit of what would amount to a diminished cost of production. It may be laid down as a general law, that in a perfectly free country, consumers will always gain in the long run by any permanent diminution in the cost of production "*pur et simple*" that is taken apart from the minimum of profit required by producers. Rough settlers in new countries can grow grain at a less cost than we in the Old World can, and if the land of England could not have been turned to any other use but that of growing grain crops, the supplies of grain poured in from foreign countries would soon have brought our rents at home down; fortunately for our farmers, they could turn their attention to the production of meat, butter and cheese, without meeting with the same kind of competition. When men farm their own land, or live in their own houses, they do not pay rent; but if they wish to know their gain or loss by the operation, they must put their farm, or their house, down in their accounts at its letting value.

256. Is it desirable to protect tenants by land laws?

It is, as long as these laws are strictly confined to securing to the tenant what justly belongs to him and no more, because they thus enable him to turn his fixed capital into movable capital, and to migrate elsewhere if he cannot agree with his landlord.

CHAPTER IX.

TAXATION.

257. What is taxation in Great Britain?

It is the cost of the British constitution, that is, the per-centage which we must pay to secure our lives and properties.

258. What ought a tax to be?

It ought to be but the price which we have to pay for some advantage given to us in return.

259. How can we best consult the interests of the whole mass of the nation?

By encouraging free competition amongst all exchangers, irrespective of race or country, and thereby lessening the cost of producing commodities.

260. Does taxation in any country necessarily restrict competition in that country?

No, it does not, provided that the same kind of commodities are taxed exactly alike, whether they be produced at home or abroad. Taxation only restricts competition when the same kind of commodities are unequally taxed, that is to say, when the price of certain kinds of commodities which some competitors supply is raised by taxation above the price of the same kind of commodities which other competitors supply.

261. What should be the aim of the financialist?

To endeavour to make competition absolutely free.

262. What are direct taxes ?

" They are taxes which are collected from house to house by Government officers, whether avoidable or unavoidable."—(Newman's " Lectures," p. 215).

263. What are indirect taxes ?

They are taxes upon the consumption of various commodities, the incidence of which can be avoided by any one who does not consume the commodities, because those who pay these taxes to the State, try to reimburse themselves by charging so much more for their goods.

264. What are the objections to direct taxation ?

That it interferes with the private concerns of the people too much, that it gives great opportunities for fraud, and that it cannot be put equally on all classes, but that those who are rich pay more than their fair share of taxation.

265. What are the objections to indirect taxation ?

That it is not just that taxation should be optional, and that any tax on the consumption of commodities must prevent the natural development of trade.

266. What is the advantage of direct taxation?

That it presses on those who are best able to bear it, and that it enables us to lighten indirect taxes on the consumption of various commodities which may be called necessaries, whereby we both improve the condition of the masses and greatly develop the trade of the country; for instance, lowering the taxes on tea, sugar, wine, &c.

267. What is the advantage of indirect taxation?

That a very large mass of the people are made contributors to the revenue, by indirect taxes, whom no system of direct taxation could reach, as in the case of the tax on tobacco in Ireland.

268. What is the best system of taxation?

A mixture of direct and indirect taxation, such as we have in Great Britain.

269. What is the proportion between these two modes of collecting our revenue, as it stands at present?

Mr. Bass asked the Chancellor of the Exchequer whether he would be good enough to state to the House what were the proportions contributed respectively by direct and indirect taxation to the gross revenue of the country (12 May, 1871).

"*The Chancellor of the Exchequer:* ' The question appears to be a very simple one, but it is really one of considerable difficulty, because of the ambiguity of the terms employed. I must, therefore, first of all, state the sense in which we employ these terms, though I am far from saying that it is the proper sense or the one in which hon. members interpret them. By gross revenue, I understand the whole revenue as estimated, including the expenses of collection. With direct and indirect taxation I have a little more difficulty, because there are seven or eight different meanings, each of which has its advocates. All taxes are in one sense direct, because they are taken from somebody by the Government, but, as I understand the terms, an indirect tax implies that the person who pays it has some means of recovering it from the community, and a direct tax implies that the person who pays it has no such power. According to this classification, the tax on tea would be an indirect tax, and the Income Tax would be a direct tax. Applying that principle, the Customs would be wholly indirect taxes, and the Excise, with some trifling exceptions, would be also indirect. With reference, however, to the Excise upon railways, I am in some difficulty, because the power of recovering it from the public depends upon very complicated considerations; for while some railways are empowered by Acts of Parliament to recover the tax from their passengers by a charge over and above the Parliamentary maximum, others have not that power. As it stands at present, we have deducted it from the Excise, and treated it as a direct tax. It yields £505,000. Then, under the head of direct taxation, stamps, less Law Court fees, amount to £8,585,900. Taxes, principally the land tax and inhabited house duty, £2,725,000, and Income Tax, £6,350,000. According to that division the direct taxation would amount to £18,165,900; and the indirect taxation to £42,744,000. There are, however, other sources of revenue which are neither direct nor indirect, and which I have excluded from this comparison on that account. These are the Law Court fees, which appear to me to be payments for services rendered, £421,100; Post Office, £4,770,000; telegraphs, £500,000; Crown lands, which are really proceeds from public property, £385,000; and miscellaneous, £3,229,200 —making in all £9,305,320. Indirect taxation, therefore, contributes about 60½ per cent.; direct taxation, 25½, and resources of revenue neither direct nor indirect, 14 per cent."

270. Why is this mixed system the best?

Because if we had only a direct system of taxation it would fall too heavily on the richer classes; direct taxes could not be collected from the poorer classes, and the great mass of the people would escape taxation altogether. If we had only an indirect system of taxation we never could have developed the trade of Great Britain to the enormous extent we have done, for we could not have lowered the taxes on such commodities as tea, sugar, wine, corn, &c., sufficiently. By mixing the two systems we are enabled to collect the required amount of revenue in the least injurious way to all classes of the nation.

271. Is the tendency in Great Britain towards direct or towards indirect taxation?

It is towards direct taxation; so much palpable benefit to the community at large has been derived from the remission of some indirect taxes, and the lowering of others.

272. How far is this tendency likely to go?

It will go on till Parliament has decided what is the highest rate which they will allow the income-tax to remain at as a permanent tax in times of peace. The system of finance which Mr. Gladstone follows has benefited enormously the mass of the people on whom the income tax does not fall, and though we perfectly agree with Mr. M'Culloch " that rich men should not be unfairly treated by being subjected to *peculiarly* high rates of taxation," we question that such men (taking the whole burden of direct and indirect taxation together) do pay a much higher per-centage on the incomes which they enjoy, than common labourers have to pay in indirect taxation alone out of their wages.

273. Would universal suffrage make the direct system of taxation go further than our present electoral system does?

It would seem in theory to be likely that the poor, who are many, would, if they had the power, make the rich, who are comparatively few, pay taxes for themselves and the poor too, but such is not found to be the case. In America, where universal suffrage obtains, the rich do not cry out that they are made to pay more than their fair share of taxation by the poor.

274. Mention Dr. Adam Smith's four rules with regard to taxation.

(1.) That the subjects of a State should contribute to the support of that State according to the revenue which they respectively enjoy under its protection.

(2.) That such contribution should not be arbitrary, both the time of payment and the sum to be paid being fixed.

(3.) That every tax should be levied at the most convenient time to the contributor.

(4.) That the expense of collection should be as small as possible.

275. If the expense of collection was precisely the same, and we wanted to raise a revenue of seventy millions by taxation, to which system ought we to resort, the direct or the indirect?

To the direct system, provided that we could levy these direct taxes equally and fairly on all her Majesty's subjects in Great Britain, rich and poor, high and low, for our trade would then be perfectly free to develop itself, and would increase enormously. Anything which cramps trade must weaken the tax-paying capacities of the people.

276. Is it just that there should be a graduated scale of income-tax, the richest people paying a

large per-centage on their incomes, and then a low per-centage gradually reached as the incomes dealt with get less?

Certainly not; such a scale interferes with the rights of property; if we are guided by strict theory, all men, whatever be their incomes, ought to pay exactly the same per-centage; those only being exempt whose incomes are so small that the tax would not pay the expense of its collection. In practice we altogether exempt incomes under £100 a year, and we allow an abatement of £80, on which income-tax is not charged, on incomes beginning with £100 a year, and ending with any sum under £300 a year; very few of the working classes, however high their wages may be, pay income-tax; while banking clerks, with incomes varying from £100 to £200 a year, have to pay, with a fourpenny income-tax, from 6s. 8d. up to £2.

277. What is the true nature of the income-tax?

It has been well defined, by Mr. Lowe, to be a tax imposed upon income strictly so called, that is, on what a man has to receive in the year, without reference to the source from which he receives it. The Income-tax is divided into five Schedules:—

Schedule A...Lands (owners), Houses, Tithes, Manors, Fines, &c.
„ B...Lands (occupiers, on half the rent).
„ C...Public Funds (British, Foreign and Colonial).

Schedule D...Trades and Professions, Railways, Canals, Mines, Iron-works, Fisheries, &c., and Foreign Property.

„ E...Public Offices (General, Local, and Railway).

278. Which of these schedules meets with the most opposition?

Schedule D, because people say that incomes derived from precarious sources ought not to be charged with the same per-centage as incomes derived from realised property. This objection, taken by itself, is a valid one; we can only meet it by taking the whole Income-tax and by pointing out its cardinal principle — that we must take no notice of the source from which the incomes taxed are derived. If we once deserted this principle we should have to strike the fundholder out of the Income-tax, because all the money for the National Debt was borrowed under Acts which contained an express provision that the debt so incurred should not be liable to any Parliamentary tax whatever. We tax the fundholder upon his income, without asking him where he gets it; that is to say, we tax him not as a fundholder, but as a citizen with a certain income. If we were to strike out Schedule D, the fundholders might say, "You are now looking to the source from which incomes are derived, and arranging the income-tax accordingly; we claim exemption, and decline to pay the tax, because certain Acts of Parliament relieve us from the obligation to do so." We may also point out to the opponents of Schedule D that the trading and commercial classes have derived great benefit from the imposition of the Income-tax, because it has enabled us to diminish taxation on the consumption of various commodities, and in this way to develop trade in all its branches to the benefit of the whole community, who come in for their share in the general prosperity (*see* 270). The truth is, that the principle of the Income-tax is involved when Schedule D is attacked, and that we cannot propose to strike it out, without running the risk of breaking up the whole scheme of the tax. When Sir Robert Peel introduced this tax, he estimated that the yield of each penny would be £700,000 a year. Ten years ago Mr. Gladstone estimated it at £1,200,000. The actual yield of 1871-72 proved to be £1,560,000; and it is estimated that the Income-tax for the ensuing year—1872-73 will yield £1,660,000.

279. How may a heavy taxation be made to appear lighter than it is in reality?

(1.) By spreading it over the greatest number of contributors possible. A certain sum must be collected for revenue; the more you can increase the number of those who must contribute towards that sum the smaller will each single contribution become, and the less grievous will the taxation appear to be.

(2.) By repealing or lowering any taxes which interfere with the free development of trade, and substituting other taxes in place of them. The repeal of the corn laws, for instance, created a large, steady trade with foreign nations, who send their grain to Great Britain, and did more good to the people by the employment caused by that trade than by the mere cheapening of the loaf. The source of wealth is trade; the more, therefore, we can develop trade in all its branches the richer we become, and the better able to pay any taxation which is found to be necessary.

280. What are excise duties?

Taxes levied upon the consumption of commodities which are produced within the kingdom.

281. What are customs duties?

Taxes levied upon the consumption of commodities which are produced in foreign countries and brought into the kingdom.

282. Is the amount of duty remitted to the consumer the only relief which he generally obtains from a remission of the custom duties levied on commodities from abroad?

No; he may gain greatly by a reduction in the price of the commodity on which the duty has been remitted, for the competition from abroad, which that remission produces, will bring down the price of the commodity, so that the consumer not only gets the relief given by the remission of the customs duty, but he also gets the relief given by the reduction in price from the increased competition. It is competition which brings the profit made by the producers of any commodity to its minimum rate, for one undersells the other till this rate is reached.

283. Give an instance to explain this.

Brandy is made in Great Britain to a very considerable extent, and is also imported from France, but paid formerly a duty of 15s. 2d. per gallon; now, when this duty is lowered to 10s. 5d. per gallon, the British manufacturer who could successfully compete with French Brandy that paid a duty of 15s. 2d. per gallon, may very likely not be able to compete with it when it pays a duty of 10s. 5d. per gallon; or if he is still able to compete must lower his price, or he would be undersold by the French manufacturer; so that in this case the consumer not only gains by a remission of the customs duty on brandy, but he gains also by a reduction in the price of the brandy which he consumes.

284. What does this show?

It shows that the customs duty and the excise duty on any one commodity ought to be equalised, in order to secure the freest competition among those who supply that commodity; if, for instance, the customs duty on foreign brandy is higher than the excise duty on home-made brandy, the distiller of brandy at home is protected against the foreign distiller, to the disadvantage of the consumer.

285. Are export duties advantageous?

If a country has almost the exclusive supply of any commodity, a moderate export duty on that commodity may be advisable, as long as it does not in the least interfere with the consumption of the commodity. If other countries also produce the commodity, an export duty on it is a bounty to encourage your rivals in the market. Experience must decide, in the first case, whether it is best to make the foreigner pay the export duty, or to make your own people pay some equivalent tax after they have got the highest price for the commodity, which would, of course, include the export duty. Export duties have not been found to answer in Great Britain.

286. Are import duties advantageous?

In no shape; they cramp the natural development of trade: but we are obliged to resort to them to get the money required for revenue.

287. If indirect taxation is absolutely necessary, what sort of commodities should we tax heavily, and what sort lightly?

We should tax heavily any commodities which are not absolutely necessary for the health and welfare of the people, but which they are in the habit of consuming, such as tobacco, gin, whiskey, brandy, etc.; but we should tax as lightly as possible such commodities as wine, beer, tea, coffee, sugar, etc., for these commodities are directly conducive to the health and welfare of the people.

288. How long ought we go on lowering a tax on the consumption of any of these last commodities?

As long as the increased consumption on account of that lowering brings the tax to the same amount at the year's end; once a commodity has reached a price which puts it practically in the hands of the people, lowering the tax on it will increase its consumption but very little.

289. State the five principles of taxation which Mr. S. Laing lays down?

(1.) Never to impose a new tax without absolute necessity, since the annoyance and dislocation of interests make even a good new tax often worse than a bad old one.

(2) Never to tax implements of trade, raw material, or other objects of necessary use in men's daily occupations.

(3) To lighten the pressure of taxation on the working classes by exempting all the main articles actually necessary to their existence, such as bread, meat, provisions, salt, clothing, candles, etc.

(4) When obliged to levy taxes on articles of general consumption, to confine high ad-valorem rates to articles such as spirits and tobacco, and for the rest to aim at raising the requisite revenue by very moderate import duties on a few leading articles, such as tea and sugar.

(5) To preserve a fair balance between direct and indirect taxation; and, speaking roughly, to divide equally between each the benefit of any reduction, and to impose on each equally the burden of any permanent increase of expenditure.

290. What is a tariff?

A table of the duties established by law in any particular country.

291. What is an ad-valorem duty?

A duty based upon the value of the commodity it is imposed on, and which fluctuates as that value fluctuates.

292. What is a specific duty?

A duty which may or may not be based on the

value of the commodity it is imposed on, but which does not fluctuate as that value fluctuates.

293. Are there, then, two kinds of specific duties?

Yes, there is the purely specific duty, which is not based on the value of the commodity it is imposed on, as for instance, the tax on tea in Great Britain of 6d. per pound, common congou, worth about 10d. per pound, paying exactly the same duty as flowery pekoe, worth 3s. 6d. per pound; and there is the specific duty which is based upon the value of the commodity it is imposed on, but which does not fluctuate as that value fluctuates. This may be called the specific duty on a classification of commodities arrived at by valuation; for instance, in the last French tariff, under the one class, linen, instead of a purely specific duty on every kind of linen, there is a specific duty of fifteen per cent. on seven sub-classes of linens, and this duty differs for each of them, as it has been arrived at by calculating the rate of fifteen per cent. on a previous valuation of each sub-class.

194. Are there any objections to the working of a tariff on the purely ad-valorem system?

Yes; such a tariff gives great opportunities for wrangling and fraud, and there must always be a danger of under-valuations to the detriment of the public revenue. The advocates of the system declare that it is the simplest, but this simplicity disappears on examination of the way the system works; it is very simple to strike an ad-valorem rate of twenty per cent. on the value of any number of commodities, but it is very far from being a simple matter to levy that rate on each particular class of commodity according to its market price.

295. Are there any objections to a tariff based upon the purely specific system?

There are no objections to the practical working of a tariff on such a system, but the difficulty of framing it successfully with regard to such commodities as wine and sugar would prevent its being universally adopted. As Mr. Fryer very justly remarked with regard to sugar, "by the removal of the treacle and dirt, the mass was reduced in weight, and, if the same duty was imposed upon the raw, with its impurities, that was charged on the pure loaf sugar, the produce of the foreign refineries, it would be clear that the British refiner would pay two duties, one upon the sugar extracted and another upon the refuse removed, while his competitor, the foreign refiner, would pay upon the former only." Since this was written the wine duties have been greatly simplified; all wine under 26 degrees of proof spirit verified by Sykes's hydrometer pays a duty of one shilling per gallon, from 26 degrees up to 42 it pays two shillings and sixpence a gallon, and for every degree beyond 42 an additional duty of three pence per gallon. The sugar duties have also been lowered, refined sugar paying a duty of six shillings per cwt., and molasses a duty of one shilling and ninepence per cwt.

296. Are there any objections to a tariff based

upon a system of specific duties on a classification of commodities arrived at by valuation?

Yes; one of the principal objections lies in the necessity of frequently revising such a tariff, in order to keep the specific duties always at the level, which they were originally designed to observe, with relation to the selling prices of the commodities. For instance, if the specific duty on any particular commodity is fifteen per cent., arrived at by a previous valuation of the commodity, should this last fall ten per cent. in price, the specific duty no longer maintains its level, but becomes a specific duty of sixteen $\frac{2}{3}$ per cent. on the selling price, and therefore requires alteration. The classification of commodities in this system too leads to disputing and fraud; for instance, in the French commercial treaty as regards cotton fabrics, they are grouped in three classes, distinguished by the width and weight of the various cloths. These groups are again subjected to sub-divisions, distinguished by the number of threads in the square inch, making in all for plain goods alone nine separate standards of tariff, each of which must be laboriously grouped out by the custom-house officers with measure, scales, and whaling-glass. The ingenuity of the exporter is stimulated by this method of fixing duties to try every means of passing his highly-taxed fabrics under a lower classification than ought to be assigned to them.

297. What is the best kind of tariff, then?

A tariff which admits all the three systems as

they can most conveniently be applied. Ad-valorem duties are the best for those goods which are subject to great and frequent variations in price. Specific duties, on a classification arrived at by valuation, offer less opportunities for fraud than ad-valorem duties, and work well if they are revised sufficiently often so as to maintain their proper level with the selling prices of the commodities they are imposed on; and purely specific duties will always recommend themselves on account of their simplicity and the security which they give against fraud.

298. What must we remember in considering the expenditure of our revenue as a nation?

That, putting out of the question the advantages which we receive in return, the mere expenditure of that revenue in the country gives hosts of our countrymen employment. This is no argument in favour of taxation; the greater the cost of production, the worse for the consumer; but it is to show, that although, taken in this point of view, all these men are unproductive labourers, they (the labourers) profit just as much by their labour as if they were productive labourers, though their employers do not. Unproductive labour is labour productive to the labourer, but unproductive to the employer. A shipwright who works in her Majesty's dockyards gets just the same wages as a shipwright who works in any private dockyard; the first is, by our case, an unproductive labourer, the second, a productive one; but each gets the same price for the labour which he has to sell. A man who has any particular kind of labour to sell, gets just the same price for it, whether it is productive or unproductive to his employer; for instance, a sailor on board a yacht gets at least as good wages as a sailor on board a trading screw; so that if the expenditure of revenue injures us, considered in the light of employers of unproductive labourers, it benefits us at the same time, considered in the light of labourers in receipt of that employment. But for this being the case, we never could bear our present taxation.

299. What is the average amount of money required for revenue in Great Britain?

£70,000,000 annually.

300. How much of this amount goes to pay the interest of the national debt?

About £26,600,000.

301. Is the rate of interest payable on this debt fixed?

No; it appears to be fixed, but it is not in reality so; the variations in the price of stock change the rate of interest constantly.

302. What is the advantage of varying the price of the stock, and fixing a nominal rate of interest?

More minute variations can take place in the price of stock than could conveniently take place in the rate of interest payable on that stock; and the more minute these variations in the price of stock are, the freer the competition amongst the capitalists who want to lend their money to Government, and the better also for the public, who can have the money they want on better terms from that very competition.

303. What influences capitalists in their offers to lend money to any Government?

They judge what a fixed annuity of three per cent. is worth in the money-market according to the market rate of interest, and then they look to the credit of the Government which is borrowing from them.

304. Are the funds equally profitable as an investment to British subjects who live at home, as to British subjects who live abroad?

No; British subjects who live abroad have only income-tax to pay on the interest derived from their investments, whereas British subjects who live at home not only have income-tax to pay on their dividends, but contribute their share as individual tax-payers towards the payment of their own interest. The £26,600,000 required to pay the interest of the national debt is all raised by taxation, and forms rather more than a third of the whole amount of the revenue collected.

305. Are the funds being high a sign of prosperity in Great Britain?

No; the funds may rise from want of confidence amongst capitalists, who buy into them because they are the safest investment.

306. What have we hitherto done towards the reduction of the National Debt?

Mr. Dudley Baxter states that the true reduction of the Debt, from 1815 to 1870, a period of fifty-five years, amounted to sixty-one millions, or at the rate of one million one hundred thousand pounds a year.

307. Is it a wise policy to impose taxes for the purpose of paying off the principal of the National Debt?

It does not appear to me to be so, though it is supported by the highest authorities in England with a very few exceptions.

All actual surplus, be it small or large, should be applied to the reduction of taxation, for it is not the amount of debt which matters, but our capabilities of punctually meeting the interest due upon it.

I, therefore, do not approve of Mr. Gladstone's plan of converting the funds under the control of the National Debt Commissioners into terminable annuities, nor of the Sinking Fund Act, still in force, which provides that any surplus revenue shall, as it accrues, be appropriated to the reduction of the National Debt.

We are getting richer each year, while the sovereigns, which our creditors must take from us by their agreement, are becoming less and less valuable from an alteration in the value of gold. Could there be two better reasons assigned for leaving the reduction of the National Debt to time? Yet there is a third to be adduced, and it is of great weight, the taxpayer can make a better use of the money extracted from him for this purpose than the Chancellor of the Exchequer can. The Budget of 1871 included an appropriation of £2,250,000 towards paying off the principal of the National Debt, and Mr. Laing asks the following question in the "Times," "Shall new and obnoxious taxes be devised, or shall the balance between direct and indirect taxation be destroyed, and a dangerous precedent set of resorting to the income-tax on every occasion for the sole and simple object of giving the Chancellor of the Exchequer £2,250,000 more to invest at $3\frac{1}{4}$ per cent. ? It seems to me, the question needs only to be stated to be answered. As a pure financial question there cannot be the least doubt that the money fructifies in the pockets of the people at a rate far higher than $3\frac{1}{4}$ per cent. To every pound taken by the tax-collector you must always add a large percentage for interference with trade, curtailment of employment and consumption, and other indirect consequences, so that you are, in fact, taking 30 shillings worth 5 per cent. to invest 20 shillings at $3\frac{1}{4}$ per cent."— ("Times," 29 April, 1871.)

I think with Mr. Gladstone, that income is in the main the proper basis of taxation, and I therefore advocate the abolition of the Probate, Legacy, and Succession Duties, which are taxes on capital. I should substitute for the Probate Duty a small graduated registration fee on all wills or letters of administration where the personal estate exceeds £100, with a heavy penalty for non-registration.

Surplus revenue arising from our increasing wealth would be far better applied to the relief of these burdens, than to the paying off a liability like the National debt, running at $3\frac{1}{4}$ per cent.

Taxing the income which the nation is able to make by the employment of its capital, may be justified, because it is a necessary evil; but taxing the capital which enables the nation to obtain any income at all cannot be justified on any grounds, and least of all on those commonly assigned, viz., that a man can die but once, and that therefore the State may on that one occasion appropriate a certain portion of his capital when it is about to be transferred to those who succeed him.

The net product of the Probate Duty in the year 1870-71 was £1,591,054 in England, £166,214 in Scotland, £98,982 in Ireland; and of the Legacy and Succession duties £2,491,871 in England, £299,047 in Scotland, £158,130 in Ireland. This forms a total of £4,805,298 withdrawn along with the expenses of collection from the capital of the nation. And why? Because certain of the citizens died.

Yet the same people who cheerfully submitted to this tax upon their capital, were also called upon, in the year 1871-72 to find £2,250,000 for the redemption of a portion of the National Debt, and twopence must be added to the Income Tax rather than suspend the operation of the terminable annuity scheme, a scheme which is thoroughly unsound, and which rests on authority alone.

308. What effect have the tremendous efforts made by the American people to pay off the principal of their National Debt by heavy taxation, had upon trade in the United States?

The Hon. David A. Wells, writing in the "North

American Review" of Boston in 1871, states the fact to be "that the people of the United States use less sugar and coffee per head than they did in 1859, and also fewer boots, shoes, hats, and other articles of necessarily universal consumption, while it is positively known that the consumption of cotton cloth measured in pounds, was less in 1870 with 39 millions of people, than in 1860 with 30 millions. The people of the United States not only buy less at home, but they also send less of these and their other home manufactures abroad than they did formerly, and what they do sell abroad, they also send in foreign ships." The erroneous financial system followed in the United States, with its protective duties and its heavy taxation for the reduction of the National Debt, has done much to strangle industry, and to paralyse trade; and the result is, that workmen dependent upon wages in the towns, can earn little or nothing above the cost of living; their wages are nominally high, from 9s. to 15s. a day for skilled men, and from 2s. 6d. to 7s. 6d. a day for unskilled men, but the power which these wages give them of obtaining the comforts of life is small.

Immigration would soon come to an end in the United States if it was not for the unceasing demand for labour which is created by the appropriation of new land; vast tracts of country, hitherto uncultivated, are only waiting for the capital necessary to reclaim them, and the labour of men's hands is a part of this capital; hence wages in the United States, or in any country similarly circumstanced, cannot fall in the long run below a certain point, as the demand for agricultural labour is always acting on the market.

309. What are Exchequer bills?

Exchequer bills are bills issued under the authority of Parliament for sums varying from £100 to £1,000, which bear interest at so much half-yearly. They form the principal part of the Unfunded Debt of the nation, and they pass from hand to hand without the necessity of a formal transfer. An option is periodically given to the holders of them to be paid their amount at par, or to exchange them for new bills, with the same advantages. The interest on such bills as are not sent in for payment, or exchanged, ceases from this period. When a certain time has elapsed from the date of their first issue, Exchequer bills may be paid to the Government at par in discharge of duties or taxes. The holder of an Exchequer bill only risks the amount of premium, which he may have paid at the time of purchase, in return for the interest which accrues during the time the bill is in his possession.

The advantage to the nation lies in the fact that Exchequer bills carry a lower rate of interest than the Funded Debt does, but this rate of interest is to a certain extent dependent upon the price of the funds. When the price of the funds is high, the interest on Exchequer bills is low, but when the funds fall, so as to become a profitable investment, the interest on Exchequer bills must be raised, or otherwise too many of them would be paid into the Exchequer, in discharge of taxes. Exchequer bills dated 11th March, 1872, carry interest at the rate of £2 10s per annum for the half-year beginning with the 11th September, and ending in March, 1873.

310. Are the post-office savings-banks in Great Britain a good institution?

They are, *provided that they are so managed as to be self-supporting*, and that we do not hear in a few years of a loss of two or three millions on their operations. By a Parliamentary return issued in November, 1868, the savings-banks had a claim on the Commissioners of the National Debt for £37,177,000, but the securities in the hands of the commissioners to meet the claim, were then worth £34,399,000; there was thus a deficiency of £2,778,000 to be made up, for which the nation was liable. There were two causes at work to produce this deficiency, first, the nature of the business undertaken for the depositors by Government, secondly, the unsound system of tampering with the deposits, which formerly prevailed. When money is plentiful and cheap, the savings-banks' deposits increase, and Government invest the deposits in consols; but when money is dear and in demand, the deposits are withdrawn for more remunerative investment elsewhere; Government must, therefore, sell out stock at a lower price, which they have bought into at a higher one, for the funds will rise when money is cheap, and fall when it is dear. The interest payable on deposits has been reduced from £3 5s. per cent. to £3 per cent., in order to prevent future loss. It is not just that the whole nation should be taxed for the special advantage of the class who invest in savings-banks.

311. Can true free trade exist when only one nation in the world is guided by its principles?

It can, but only amongst the individuals who compose that nation; it does not exist with regard to her foreign trade. This one nation has herself permitted free competition, irrespective of race or country, in her own markets; but she is prevented by the protective laws of other nations from competing freely in their markets. Unless there is the same free competition in the markets on each side, there is no true free trade between nation and nation; there is only an approximation to it on the part of one nation.

312. If we permit other nations to compete freely in our markets, so as to enable them to sell to us and to buy of us, as it suits their advantage, what ought we to expect in return?

To be allowed to compete freely in their markets; or if they put import duties on our manufactures; to be allowed to buy raw material from them either free of export duty altogether, or subject to so moderate a duty that it will not interfere with our consumption of that material.

313. What has been the policy of Great Britain hitherto with regard to the commercial restrictions imposed by foreign states?

It has been to disregard such restrictions altogether, and act as if they did not exist. English statesmen consider that their country would lose more by a retaliatory system than she could gain.

314. Give an instance where a counterbalancing import duty has not been considered advantageous to English interests.

There are the paper-manufacturers in England. Rags are the only materials of which paper of a sufficiently good quality for printing can be made with profit. They are very scarce in England, as American agents buy them largely for the United States. The French and Germans have now free access to the English markets, which are the best in the world, for the sale of their paper; they have an abundant supply of rags, which they maintain at an artificially low price at home by putting a heavy export duty on them, in order to be able to distance the English paper-manufacturers in their own markets. The English manufacturers said, "If the French and Germans will charge us no export duty on their rags, let them send as much paper as they please to English markets; but it is very hard upon us that they should keep their rags at an artificially low price at home for the avowed purpose of gaining an advantage over us in our own markets. What we ask for, then, is such an import duty on French and German paper manufactured of rags, as will exactly counterbalance the export duty which the French and Germans charge us on the rags themselves." The British nation answered, "We admit that your case is a hard one, but we cannot tax ourselves by a counterbalancing duty, which must raise the price of paper, for your sole advantage; we should lose more than we could gain by doing so."

315. What, then, is the nature of the question with regard to counterbalancing import duties?

A question of pounds, shillings, and pence. Do we lose more by the rise in price of the commodity on which we put the counterbalancing import duty than we gain by taxing ourselves for the advantage of a particular class? In nine cases out of ten we lose more than we could gain; and therefore counterbalancing import duties will be very rarely resorted to in England.

316. What is the principle that nations who maintain their raw material at an artificially low price at home, by placing export duties on it, most probably act on?

It is that there is less profit to be made on the sale of mere raw material than there would be on the goods manufactured of that raw material; and that the additional profit to be made by the rise in price of raw material, which would take place from all the world having free access to it, would by no means counterbalance the serious competition which their own manufacturers must then meet with, for all other manufacturers would be put on much more equal terms with them.

317. What effect have export duties which tend to keep any raw material at an artificially low price at home?

They have the effect of bounties to the home manufacturers who require that raw material.

318. What, then, is the nature of the question with regard to such export duties?

A pounds, shillings and pence question; which is it best worth a nation's while, to let its raw material rise to what all the world will give for it, and boldly meet the competition of the world in its manufactures, or to keep its raw material at an artificially low price by putting export duties on it, and thus give a bounty to its own manufacturers, and enable them to enter those markets which they are allowed free access to with peculiar ad-advantages?

319. What system is followed in Canada at present with regard to import duties on manufactured goods?

The Canadian Government gives bounties to Canadian manufacturers, as far as it can, by putting heavy import duties on manufactured goods coming from other countries.

320. Is this system followed in the United States?

Unfortunately it is; but in spite of the heavy import duties, our manufacturers are often able to compete for a time with the American manufacturers from the very high price of goods.

321. Compare the doctrines of the protectionist and the free-trader.

The protectionist looks upon each member of the community as a producer, and says to him, " I will arrange transactions so as to make your produce sell as high as I can, and then you must be prosperous, because you are certain to get the highest possible price for what you have to sell.

The free-trader says, "We are all consumers, and therefore if we try by perfectly free competition to lower the price of all produce, we must all benefit by doing so, for we thus encourage the consumption of commodities and increase the demand for labour.

"Mr. Wells states that in 1860 there were employed in New York city alone 15,000 men in building and repairing marine steam engines, while in 1870 less than 700 men found employment in this, which is one of the highest and best paid branches of American industry, and one in which American artisans formerly excelled. Yet this destruction of a business of which the nation was justly proud has happened, he says, in the face of a rise of wages in the same industry in England. American investigators of iron shipbuilding in Great Britain report that since 1863-4 wages in that trade have advanced about 15 per cent., but, notwithstanding this, owing to the use and improvement of new machinery and the better application of knowledge, the cost of construction has declined; and from this Mr. Wells draws the inference that the result of the last ten years in the United States has been to decrease the purchasing power of wages, increase the cost of the manufactured product, diminish consumption, and prevent exports, while in Great Britain the result has been an increase of wages, a decreased cost of the finished product, an increase of consumption, and a large augmentation of exports."—(*Times*' Correspondent in the United States.)

322. Does protection protect the producer?

We give another extract from the same excellent correspondence in the "Times" as the best answer to this question, now so often asked in the United States:

"Of the bad effect of protection upon the industries it is designed to protect, Mr. Wells gives an instance in the manufacture of fur and felt hats. This industry was established in America before the Revolution, and was then so prosperous that the English Parliament passed measures hostile to it. Previously to 1860 the United States made better and cheaper hats than any other country. A seventh part of all her product was exported. A machine of ingenious character had been invented which shaped and formed the hat almost automatically. The condition of the business now is that the exportation of hats is diminished. Nova Scotia, the West Indies, Australia, and the Cape of Good Hope, which formerly bought hats of the United States, now get them elsewhere; and the price here has so far increased that the people wear proportionately fewer hats than formerly. The business, in truth, has lost its prosperity, and within the last two years the leading American manufacturers and dealers have suffered immense losses or become bankrupt. Mr. Wells regards the reason of this as so obvious that he says, no one who will take pains to examine the question can possibly miss it. The body of the hat is composed of fur or wool, either separate or mixed. We import coney fur from Germany, and if imported on the skin it pays 10 per cent. duty, and if cut from the skin 20 per cent. The reason of this difference of duty is found in the fact that but one very prominent firm in the United States cuts hatters' fur. They are said to have a machine that does the work, with small manual labour—a machine that has never been patented, and is guarded with the utmost secrecy, for fear of imitation and use either here or abroad. The parties interested have made an immense fortune out of the business, and desire that their successors shall do likewise. If wool is used instead of fur, experience shows that the most desirable, on account of its peculiar felting qualities, is wool grown at the Cape of Good Hope. Upon this the manufacturer pays an import duty of about 100 per cent. The inside silk lining, a speciality of silk imported from France, pays 60 per cent. The silk riband for the outside pays 65 per cent., while the inside leather or "sweat band," pays 45 per cent. The hat itself, if manufactured in Europe, from fur and other materials, entirely free from all these taxes, is admitted into this country at 35 per cent., and if it is made from wool the duty is 20c. to 50c. per lb. and 35 per cent. 'Is it any wonder,' asks Mr. Wells, 'that, under these circumstances, the hat business does not flourish in the United States, and that our people pay more for hats than the people of any other country on the face of the globe?'"—(*Times*' Correspondent.)

323. State why such importance should be attached to improving the education of the people?

Because this is the best way of promoting habits of self-reliance, thrift, and independence amongst the poorer classes. "The great resources of the labouring class for their happiness," says Mr. Malthus, "must be in those prudential habits, which, if properly exercised, are capable of securing to them a fair proportion of the necessaries and conveniences of life from the earliest stages of society to the latest." The common error with regard to population, which the Roman Catholic Church has supported by all the weight of its teaching, is that men should increase and multiply without regard to the means of subsistence—that population, in short, taken by itself, is a source of wealth to the State.

"Lorsqu'on s'imagine," says M. Rossi, "que la propagation de l'espèce humaine est un fait sur lequel la Providence, par je ne sais quelle exception, dispense l'homme de toute réflexion et de toute prévoyance, lorsqu'on croit que notre race n'a rien de mieux à faire que de tasser sur la surface du globe comme l'herbe des prairies, il est logique de se représenter avec délices chaque partie de terre occupée par un homme, qui n'aurait d'autre occupation que celle de lui arracher à coups de bêche sa nourriture journalière. Mais après s'être extasié sur le bonheur ineffable de tous ces hommes réduits à une faible ration de pommes de terre ou de maïs, il faut être logique jusqu'au bout et reconnaître que chez un peuple ainsi constitué, il faudrait désespérer de tout progrès de la richesse nationale ; on ne pourrait pas même garantir ces *bancs* d'hommes du retour périodique des plus épouvantables disettes." ("Cours d'Economie Politique").

We cannot make men prudent by act of Parliament, and population must be left to regulate itself, but our only hope of counteracting the natural tendency to increase and multiply, implanted in a man's mind, is to plant alongside of it, in his earliest youth, a desire to better his condition, and a determination not to do without what respectable men in his own class find necessary to their comfort. L'aisance des classes inférieures n'est donc point incompatible, ainsi qu'on l'a trop répété, avec l'existence du corps social. Un cordonnier peut faire des souliers aussi bien dans une chambre chauffée, vêtu d'un bon habit, lorsqu'il est bien nourri et qu'il nourrit bien ses enfants que lorsqu'il travaille, transi de froid, dans une échoppe au coin d'une rue. On ne travaille pas moins, ni plus mal, quand on jouit des commodités raisonnables de la vie. Le linge est aussi bien blanchi en Angleterre, où les blanchisseurs font leur métier commodément dans leurs maisons, et ne sont pas forcés de l'aller péniblement savonner à la rivière,—(Jean Baptiste Say).

324. Is there any other way of promoting the independence of the working classes ?

We may do much more than we now do in this direction, by giving as little assistance from the public purse as we possibly can to those who do nothing to help themselves Mr. W. H. Smith, M.P., made the following remarkable statement in Parliament, in May, 1871:—

"Side by side with our increasing wealth and commercial prosperity, pauperism was also increasing to an alarming degree, and constituted a problem which deserved the most serious consideration of the House and the country. He would for a moment recur to figures showing the growth of the metropolis during the last ten years. In 1860 the population of the metropolis was estimated at 2,770,000, the number of paupers was 86,000, and the amount expended in their relief was £796,000, or a burden of 6s. 4d. per head of the population. In 1863 the population had increased to 2,904,000, the paupers to 94,000, and the expenditure to £868,000, or 6s. 7d. per head of the population. In 1867 the population stood at 3,082,000, the paupers at 126,000, and the expenditure had gone up to £1,175,000, or 8s. 8d. per head. In the year 1870—and hon. members would now perceive the full force of the contrast between the two periods—the population was taken at 3,215,000, the pauper class at 141,000, and the expenditure in poor relief had reached £1,466,000, or an average burden of 9s. 1d. per head upon the whole population of the metropolis. These figures showed that the cost of administering the Poor Law of the metropolis had increased by 84 per cent. within ten years; there had been an increase in the number of outdoor paupers to the extent of 77 per cent., the indoor paupers were 35 per cent. more numerous in 1870 than in 1860, but the population generally had only increased by 16 per cent. during the same period of time. But if the increase in the pauperism and expenditure of the metropolis, alarming though it was, were compared with that of the country generally, the contrast became more alarming still. The population of the country, exclusive of the metropolis, might be taken at 17,000,000; and, while the paupers had increased in number from 717,000 to 827,000, the expenditure had gone up from £4,600,000 to £6,200,000, representing an advance of about a third, while the expenditure of the metropolis had nearly doubled in the same period. The burden imposed upon the population for the maintenance of paupers was 9s. 1d. per head per annum in London against 6s. 7d. in the remainder of the country. He found that in Mile-end the charge for indoor relief had increased from £5,442 in 1860 to £6,200 in 1870, the outdoor relief in the same period having increased from £3,370 to £8,089. In Poplar the indoor relief was £6,054 in 1860, and £9,172 in 1870, the outdoor relief being £9,524 in 1860, and £23,348 in 1870. As there had been in Poplar distress arising from exceptional causes, he should not insist strongly upon that instance. In Hackney, again, the cost of indoor relief was £4,505 in 1860, and £7,730 in 1870, and of outdoor relief, £4,978 in 1860, and £19,864 in 1870. As the result of personal inquiry, he found that the class of persons who were demoralized by the receipt of parochial relief were not instructed artisans, but persons of the uninstructed labouring classes, who had no regular employment, but earned what they could, and spent it as it came, with result that if they were out of work at the end of a week they went for assistance to the parish on the following Monday morning. One of two things was clear from this state of facts, either that these classes were insufficiently paid or their habits were careless and improvident in a remarkable degree. It was also true that the classes who were least certain in their earnings or employment were the persons who most readily married and took upon themselves without thought the responsibility of maintaining families."

The administration of out-door relief in London was doing most serious mischief until it received a check from the passing of an Act which throws the cost of out-door relief upon the particular parish which gives it, while the indoor relief is paid by a common rate. Cases where out-door relief is given ought to be very carefully inquired into first, because experience shows that out-door relief has the effect of making certain classes dependent, and of destroying their feelings of self-reliance—they look to parochial aid at once if they get any encouragement to do so.

Let us hope that each year will convince our poor-law guardians more and more of the vast importance which attaches itself to the manner in which they perform their duties; they are merciful, but they must be strict too. I add a return, showing one of the effects of our present prosperity:—

"The first half of the year 1872 duty was paid on 12,282,478 gallons of home-made spirits for consumption, as beverage only, in the United Kingdom, being an increase of 1,142,514 gallons over the corresponding half of the preceding year. The quantity retained for consumption in England was 6,646,059 gallons, an increase of 625,996 gallons; in Scotland, 2,821,468 gallons, an increase of 324,867 gallons; in Ireland, 2,814,951 gallons, an increase of 191,651 gallons. The quantity of foreign spirits entered for consumption in the United Kingdom in the half-year was 4,117,345 proof gallons, an increase of 15,194 gallons over the corresponding half of the preceding year."

325. What does the science of which we treat teach us with regard to the management of our revenue in Great Britain?

It does not pretend to teach us what expenditure ought to be maintained, but it teaches that we should have a proper return for that expenditure, whatever it be.

326. What is the Budget?

The statement which the Chancellor of the Exchequer makes each year concerning the public income and expenditure for the financial year which is to come is called the Budget, derived from the French word "bougette," a small leather bag. We give the Budget estimate for the financial year 1872-73:—

PUBLIC INCOME AND EXPENDITURE.

	£			£
1. Customs	20,080,000	1. Charges on Consolidated Fund—		
2. Excise	23,310,000	Funded and Unfunded Debt	£26,830,000	
3. Stamps	9,700,000	Other Charges	1,780,000	28,610,000
4. Land Tax and House Duty	2,300,000	2. Supply Services—		
5. Income Tax	6,940,000	Army	£14,824,000	
6. Post Office	4,770,000	Abolition of Purchase	853,000	
7. Telegraph Service	850,000	Navy	9,508,000	
8. Crown Lands	375,000	Miscellaneous Civil Service	10,652,000	
9. Miscellaneous	3,300,000	Revenue Department	2,621,000	
		Post Office	2,610,000	
		Telegraphs	500,000	
		Packet Service	1,135,000	42,703,000
		3. Surplus of Estimated Income over Estimated Expenditure		312,000
	£71,625,000			£71,625,000

TREASURY, *October 1st, 1872.*

There are three items on the expenditure side. No. 1 consists of the charges on the Consolidated Fund, a certain part of the public revenue set apart each year to meet these charges. They are not the subject of a vote in the House of Commons, as they must be paid without further question once they are put on the Consolidated Fund.

These charges on the Consolidated Fund are again divided into two heads, the expenses connected with the management of the Funded and Unfunded Debt, and "the other charges."

This last item is again divided thus :—

>Civil List,
>Annuities and Pensions,
>Salaries and Allowances,
>Diplomatic Salaries and Pensions,
>Courts of Justice,
>Miscellaneous charges.

The Civil List is the sum allotted to Her Majesty the Queen, and amounts to £385,000. The Privy Purse, consisting of £60,000, is the only part of this sum over which Her Majesty has absolute control and disposal; the balance, or £325,000 is parcelled out amongst the various departments of the household.

When we compare the expense of our Sovereign power with the expense which other great nations incur for the same object, we have no reason to complain of our present system, and we may be grateful to Queen Victoria, who acts on the good old principle " out of debt out of danger."

No. 2 consists of the Supply Services which are voted each year by the House of Commons.

We now expend above 24 millions on our army and navy; but if we look back to 1853, we find Mr. Gladstone, on the 18th April, estimating the cost of these services at £12,860,000 for the financial year 1853-54. Fortunately for us, the revenue, exclusive of the income-tax is augmenting at the rate of £3,500,000 a year, and we may be thankful for the improvement which Sir Robert Peel and Mr. Gladstone have effected in our financial system. There is still much to be done in this direction. We ought to apply all surplus to the reduction of taxation, and to make no attempt to pay off the National Debt, either by Sinking-fund Acts or by Terminable Annuity schemes.

No. 3 is the surplus which will be very much larger than the estimated one, but as we submitted ourselves to be governed by an "ex post facto" law in the Alabama case, and as we were justly condemned by fair arbitration to pay a fine because we did so submit ourselves, our surplus will be sent to New York.

When we examine the income side of the account, we are struck by the simplicity of its arrangement: sugar, tea, tobacco, spirits, wine, beer and ale, malt, coffee, cocoa, preserved fruits, and a few other minor heads are the only articles of consumption which we impose taxes on. We owe this simplification, and the vast development of trade which has resulted from it, to the introduction of the income-tax as a permanent source of revenue. Let us hope that we may always have the good sense to continue it. We extract the following account of the public income for the year ending 30th March, 1872 from the "Times" of 3rd October, 1872 :—

"The Finance Accounts recently issued for the financial year ending the 31st of March, 1872, show that the revenue received into the Exchequer amounted to £74,708,314. The 'Statistical Abstract,' which extends over the last 32 years, shows also that in all that time this amount of revenue has only once been exceeded—viz., in the financial year 1869-70. The following statement explains how the revenue of the year 1871-72 was obtained. The Customs' duties produced £20,326,000. The chief items are £6,797,018, the amount of the duties levied on the import of tobacco and snuff into the United Kingdom ; £4,523,848 on the import of foreign and colonial spirits; £3,179,930 on sugar; £3,079,284 on tea; £1,646,735 on wine. Four of these items are constantly increasing in productiveness, but the duty on sugar, which was largely reduced in 1870, produced above £6,000,000 sterling ten years ago. The excise duties produced the large amount of £23,326,000 in the year 1871-72. The largest item, an ever-increasing item, is £12,274,596 levied on British and Irish spirits; the duty on malt brought in £6,670,955 ; the railway passenger tax, always increasing in productiveness, £527,567. The item of licences is becoming very large, and reached £3,781,979. Among them are these:—£753,229 for spirit licences granted to distillers, dealers, and publicans ; £428,250 for licences to brewers; £368,496 licences for sale of beer, cider, and perry; £524,206 licences for carriages; £433,277 for horses; £279,425 for dogs; £198,946 for male servants; £193,843 for sale of wine and sweets; £189,824 for game certificates or licences ; £62,437 for gun licences; £86,652 for armorial bearings; £64,228 for auctioneers and appraisers ; £35,032 for pawnbrokers. The next class of duties is that of stamps, which produced in the year 1871-72 no less than £9,772,000, which is a larger amount than in any other year in the 32 years' list. The duty on legacies and successions produced £3,360,489, and on probates and administrations, £1,852,560; the latter amount slightly exceeded in the preceding year, the former never equalled. The following amounts are very large :— Deeds, etc., £1,842,422 ; bills of exchange, £846,258 ; receipts, draughts, and other penny stamps, £648,843. The item of stamps used for the collection of fees in courts of justice and public offices reaches £525,788. The next class of taxation retains the imposing title of 'Taxes,' but now includes only the land-tax, amounting in the year 1871-72 to £1,086,350, and the inhabited house duty, which shows an increase to £1,262,611; but the account of payments into the Exchequer states these taxes at only £2,330,000. The income-tax also is stated at £9,084,000, but the detailed statement puts the amount of this tax for the year at £9,328,102, and the amount is approximately estimated as appropriated thus :—Schedule A, £3,325,492 ; B, £434,903; C, £897,541; D, £4,125,324; E, £544,842. This tax was at 6d. in the pound. These five items—Customs, Excise, stamps, income-tax, and land-tax and house duty—constitute our taxation proper, amounting to about £65,000,000 sterling in the year 1871-72. Then comes the Post Office, producing a payment of £4,680,000 into the Exchequer, not deducting the expense of conducting the service ; and the telegraph service, with £1,183,156 gross receipts and £755,000 net produce. The Crown lands supplied £375,000. The account closes with the large item of 'Miscellaneous receipts £4,060,314.' It includes fourteen classes of receipts. The first comprises small branches of the hereditary revenue, £25,393. Next comes £138,578, payable by the Bank of England out of profits of issue. Then, fees received in public offices, £569,396, including an item of £302,972 from County Courts. Receipts by naval and military departments come next, amounting to £1,177,794—viz., contributions from Indian and Colonial Governments and sales of old stores. There is also a further sum of £496,842, contributed by India for military and diplomatic charges. Next stands £380,656 receipts by Civil departments, including £53,293 for convict establishments, chiefly repayments of expenditure for Colonial Governments ; also £45,049 repayment for extra constabulary force in certain Irish counties ; £35,608 from Dublin, chiefly police tax; £17,659 proceeds of sales of books at the Irish Education Office ; £45,806 for seigniorage on silver purchased for coinage ; £34,910 proceeds of the sale of books and waste paper at the Stationery Office ; £39,348 sale of old materials by the Office of Works and receipts from Parks, etc. Next comes £332,419, receipts by revenue departments for various services, including repayments from India and Australia on account of the mail service, Custom House charges on delivery of goods from bonded warehouses, inland revenue fines, and charges of management of Post Office banks repaid by the National Debt Commissioners. Then comes £13,440 the income of the *Gazettes;* then £883,144 over-issues repaid, chiefly the Abyssinian Grant. The rent of the Malta and Alexandria telegraph, £16,320, follows, and then £1,918 profit on Treasury chest transactions. Customs revenue and repayments of expenditure amounting to £12,458 come from the Isle of Man, and £8,000 from Greece, towards the sum paid by us under our guarantee of the Greek Loan. The list closes with £3,956 casual receipts; among them is £630 for fines inflicted by naval authorities on African chiefs for breach of treaties. Thus was our year's revenue of nearly 75,000,000 sterling made up."

327. What should we desire most at present for the general advantage of all classes in England?

That the nations we deal with should become free traders like ourselves, for we should then have plenty of employment at home, and consequently be better able to pay whatever taxes are found to be necessary.

CONCLUSION.

We judge of truth in practical matters from facts and from life, for on them the decisive point turns; and we ought to try all that has been said by applying it to facts and to life; and if our arguments agree with facts, we may receive them, but if they are at variance, we must consider them *as mere words.*

ARISTOTLE'S "ETHICS," Book 10, Chap. 8.

www.ingramcontent.com/pod-product-compliance
Lightning Source LLC
Chambersburg PA
CBHW031826230426
43669CB00009B/1240